The Best American Science and Nature Writing 2012

GUEST EDITORS OF
THE BEST AMERICAN SCIENCE
AND NATURE WRITING

2000 DAVID QUAMMEN
2001 EDWARD O. WILSON
2002 NATALIE ANGIER
2003 RICHARD DAWKINS
2004 STEVEN PINKER
2005 JONATHAN WEINER
2006 BRIAN GREENE
2007 RICHARD PRESTON
2008 JEROME GROOPMAN
2009 ELIZABETH KOLBERT
2010 FREEMAN DYSON
2011 MARY ROACH
2012 DAN ARIELY

The Best American Science and Nature Writing™ 2012

Edited and with an Introduction
by Dan Ariely

Tim Folger, Series Editor

A MARINER ORIGINAL
HOUGHTON MIFFLIN HARCOURT
BOSTON • NEW YORK 2012

www.hmhbooks.com

ISSN 1530-1508
ISBN 978-0-547-79953-7

Printed in the United States of America

DOC 10 9 8 7 6 5 4 3 2

"The Brain on Trial" by David Eagleman. First published in *The Atlantic*, July/August 2011. Copyright © 2011 by David Eagleman. Reprinted by permission of the author.

"Dream Machine" by Rivka Galchen. First published in *The New Yorker*, May 2, 2011. Copyright © 2011 by Rivka Galchen. Reprinted by permission of Rivka Galchen.

"The Feedback Loop" by Thomas Goetz. First published in *Wired*, July 2011. Copyright © 2011 Condé Nast. All rights reserved. Reprinted by permission.

"The Peanut Puzzle" by Jerome Groopman. First published in *The New Yorker*, February 7, 2011. Copyright © 2011 by Jerome Groopman. Reprinted by permission of Jerome Groopman.

"How to Hatch a Dinosaur" by Thomas Hayden. First published in *Wired*, October 2011. Copyright © 2011 Condé Nast. All rights reserved. Reprinted by permission.

"Our Body the Ecosystem" by Virginia Hughes. First published in *Popular Science*, March 2011. Copyright © 2011 by *Popular Science*. Reprinted by permission of Bonnier Corporation.

"Ill Wind" by David Kirby. First published in *Discover*, April 2011. Copyright © 2011 by David Kirby. Reprinted by permission of David Kirby.

"Sleeping with the Enemy" by Elizabeth Kolbert. First published in *The New Yorker*, August 15 & 22, 2011. Copyright © 2011 by Elizabeth Kolbert. Reprinted by permission of Elizabeth Kolbert.

"The City Solution" by Robert Kunzig. First published in *National Geographic*, December 2011. Copyright © 2011 by National Geographic Society. Reprinted by permission.

"Mad Science" by Mark McClusky. First published in *Wired*, March 2011. Copyright © 2011 Condé Nast. All rights reserved. Reprinted by permission.

"Ants & the Art of War" by Mark Moffett. First published in *Scientific American*, December 2011. Used with permission. Copyright © 2011 by *Scientific American*, a division of Nature America, Inc. All rights reserved.

"Deep Intellect" by Sy Montgomery. First published in *Orion*, November/December 2011. Copyright © 2011 by Sy Montgomery. Reprinted by permission of Sy Montgomery.

"The Touchy-Feely (but Totally Scientific!) Methods of Wallace J. Nichols" by Michael Roberts. First published in *Outside*, December 2011. Copyright © 2011 by Michael Roberts. Reprinted by permission of Michael Roberts.

"Crush Point" by John Seabrook. First published in *The New Yorker*, February 7, 2011. Copyright © 2011 by John Seabrook. Reprinted by permission of John Seabrook.

"Test-Tube Burgers" by Michael Specter. First published in *The New Yorker*, May 23, 2011. Copyright © 2011 by Michael Specter. Reprinted by permission of Michael Specter.

"The Wipeout Gene" by Bijal Trivedi. First published in *Scientific American*, November 2011. Copyright © 2011 by *Scientific American*, a Division of Nature America, Inc. All rights reserved. Reprinted by permission of Bijal Trivedi.

"The Long, Curious, Extravagant Evolution of Feathers" by Carl Zimmer. First published in *National Geographic*, February 2011. Copyright © 2011 by Carl Zimmer. Reprinted by permission of Carl Zimmer.

Contents

Part Three: Humans (the Good)

Part Four: Humans (the Bad)

Part Five: Society and Environment

Part Six: Technology

Foreword

LAST NOVEMBER, WHILE gathering articles for this collection, I read that Lynn Margulis had died. She was one of the great evolutionary biologists of our time, and exceedingly controversial. In 1967, when she was twenty-nine, she published a paper that transformed our understanding of the evolution of life. Before being accepted by the *Journal of Theoretical Biology*, the manuscript had been rejected by fifteen other journals—the academic equivalent of publishing houses turning down the first Harry Potter book. Her forty-nine-page article challenged one of the bedrock principles of modern biology: that random mutation was the prime driver of evolution. Margulis argued instead that some of the most crucial evolutionary developments in the 3.8-billion-year history of life on Earth were the result of cooperative and mutually beneficial relationships among organisms. Specifically, she was convinced that more complex forms of life arose when simpler ones merged into a single organism—to the advantage of all parties involved. She called the process symbiogenesis.

Margulis looked to the humblest living things to find evidence for her hypothesis. She argued that in the distant past, primitive single-celled creatures combined, creating more elaborate cells that would eventually give rise to all higher forms of life. The traces of those ancient unions remain today in nearly every cell of our bodies. It is now generally accepted that mitochondria—microscopic components of our cells that provide the chemical energy that keeps us alive—once existed as free-living bacteria that were engulfed by some larger cell, an event that probably happened roughly 2 billion years ago. Chloroplasts, the tiny engines of photosynthesis found in all plant cells, had similar origins. Without symbiogenesis, there would be no flowers, no trees, no grass, no animals, insects, or people. There would be no oxygen in the atmosphere.

We humans have a self-important view of our place in the world. "Evolutionists have been preoccupied with the history of animal life in the last 500 million years," Margulis once wrote. She championed a different perspective; she was a defender of the microscopic beings that make all other life possible. Bacteria were the only life on Earth for about half the planet's history. In a sense, they remain dominant even today, thriving in just about any habitat, from glacial ice to blistering hot springs. The total mass of the world's bacteria is probably greater than that of all other life combined. Human bodies are prime bacterial real estate. Our intestines carry several pounds of bacteria—the number of individual *E. coli* in the guts of each of us exceeds the number of humans who have ever lived. Be thankful for their presence—we couldn't digest food without them.

Two articles in this collection provide a salutary Margulisian corrective to our self-regard. Brendan Buhler, in "The Teeming Metropolis of You," begins his story with a wonderful sentence: "You are mostly not you." Our vast colonies of resident bacteria, Buhler tells us, protect us from disease and may even influence the development of our brain. You might then be primed to read "Our Body the Ecosystem," by Virginia Hughes, which describes how some inflammatory skin conditions, such as eczema and psoriasis, may be caused by imbalances among the bacterial populations that live on us.

Dan Ariely, this year's guest editor, has arranged his selections in what he referred to in a note to me as a "micro to macro" order, beginning with three stories about bacteria and ending with five on technology. It's an arrangement, I think, that reflects another aspect of Margulis's thinking: that Earth itself can be regarded as an enormous example of symbiosis. Forty years ago, she and James Lovelock, an independent British scientist, collaborated on the Gaia hypothesis, which likens Earth and its web of ecosystems to a living organism. Depending on your point of view, Gaia is either a romantic metaphor or a provocative scientific theory. Or perhaps it's both.

I wish this book had more pages. As always, the range of topics is extraordinary, as you can see from a glance at the table of contents: a history of feathers; a scheme to re-create a dinosaur; ant wars; our relation to Neanderthals; the mysteries of teenage behavior; the inner lives of octopuses; and other topics that Dan describes in his introduction. And, as always, there just isn't enough room for all the stories that deserve a place here. Among them is an interview with Lynn Margulis by Dick Teresi. ("I don't consider my ideas controversial," she told Teresi, "I consider them right.") You can find the interview in the April 2011 issue of *Discover;* it is also available online. So ease into a comfortable chair and let the trillion-celled colonial organism that is you enjoy some of the best writing and reporting you'll find in any genre.

I hope that readers, writers, and editors will nominate their favorite articles for next year's anthology at http://timfolger.net/forums. The criteria for submissions and deadlines, and the address to which entries should be sent, can be found in the "news and announcements" forum on my web site. Once again this year I'm offering an incentive to enlist readers to scour the nation in search of good science and nature writing: send me an article that I haven't found, and if the article makes it into the anthology, I'll mail you a free copy of next year's edition. I'll even sign it, which will augment its value immeasurably. (A true statement, by the way, there being no measurable difference between copies signed or unsigned by me.) Perhaps I'll manage to cajole Dan into signing some copies. I also encourage readers to use the forums to leave feedback about this collection and to discuss all things scientific. The best way for a publication to guarantee that its articles are considered for inclusion in the anthology is to place me on the subscription list, using the address posted in the news and announcements section.

It has been a pleasure to work with Dan Ariely. I'm looking forward to reading his latest book, *The Honest Truth About Dishonesty.* Once again this year I'm indebted to Amanda Cook and Ashley Gilliam at Houghton Mifflin Harcourt. I would also like to dedicate this year's anthology to my mother, with the hope that she will be reading these words when the book is published in October. And I will forever be indebted to my beauteous wife, Anne Nolan.

TIM FOLGER

Introduction

On Science-Based Paternalism

SCIENTIFIC PURSUITS ARE constantly uncovering the inner workings of our psyches and our worlds, from the quantum to the cosmic level. In this volume, you will find essays that discuss the biota that fills and surrounds our bodies; animals that mimic and inform our own behavior; human failures and strengths; the societies and environments we create and inhabit (and sometimes destroy); and the technologies we develop and depend on. And although the journey from microcosm to macrocosm may seem like explorations of very different worlds, you will see that a common thread runs through the articles. Each of them offers an independent but complementary musing on the human condition. (Of course, I may be biased in this interpretation, given my own research focus.) To me, the main theme that surfaces in all of these articles is the timeless question of what makes us human. Is it the microbes that constitute the bulk of the cells in our bodies? Our differences from or similarities to other species in the animal kingdom? Our ability or irrepressible impulse to manipulate our environments with a godlike hubris? Our genes, our brains, the pheromones we emit? Or, as Elizabeth Kolbert suggests in "Sleeping with the Enemy," is it our unique sense of *Homo sapiens* madness? The main conclusion I've reached, based on the evolution of science in general—and as reflected in this collection—is that we are extraordinary yet flawed and predictably irrational creatures.

Contained within each of these articles, and with each new scientific discovery, is the joy of learning for its own sake, which is considerable in itself. But research is only the first step; we can't ignore the lingering question of how we can—and should—use scientific evidence to our advantage. How do we make the most of science so that it has a greater purpose than merely quenching our curiosities?

I have always looked at science as a tool to improve our lives and environment; it's why I got into the field in the first place. When I was in the hospital many years ago being treated for extensive burns, the nurses would rip the bandages off my skin grafts as quickly as possible. They figured, following conventional wisdom, that short bursts of pain were preferable to prolonged (albeit less intense) periods of pain. Once I was out of the hospital and enrolled in Tel Aviv University, I turned my first academic research efforts toward figuring out whether people suffered less when pain was brief but intense or longer-lasting but moderate. As it turned out, people fared better with the latter. Reflecting on the gap between how my kind and highly experienced nurses had treated me and what I found in many controlled experiments showed me that my nurses hadn't figured out the best way to remove bandages. Beyond the issues of burn patients, this made me wonder if we rely on biases and inaccurate beliefs about how the world works across many aspects of life to the detriment of ourselves and others. We think we are rational creatures, but time and time again we misjudge our own preferences, make decisions that fall far from our best interests, and are swayed by forces that we don't recognize.

What this suggests is that there is ample room for improvement in almost any aspect of human enterprise and that basing our decisions, guidelines, and procedures on scientific findings is the right way (and, in my opinion, the only way) to move forward. This is the upside of using the scientific method to inform our day-to-day actions. As Thomas Hayden quotes in "How to Hatch a Dinosaur," "What good is a toolkit if you don't use it to build something?"

As we assemble a scientific toolkit of information, we can start remodeling the world around us to build stronger reinforcements that support our multifaceted human nature. We can take what we are learning about the microscopic world of gut biota to develop probiotics to treat conditions ranging from eczema to allergies. Or use the perpetually evolving science of genetic engineering to combat issues from obesity to dengue fever. And if we find that people are more likely to behave in environmentally friendly ways when they feel emotionally connected to nature, we can try out new methods of encouraging conservation that don't use guilt tactics or information overload. We can test new strategies that promote sustainable behavioral change, and when we figure out the best ways to reinforce positive outcomes, we can enlist them on a larger scale.

Scientific findings do not always portray us in a favorable light. Although we sometimes discover how wonderful we are or how difficult it is for computers to reason the way we do so naturally (see "Mind vs. Machine" by Brian Christian), we occasionally uncover things about ourselves that are not particularly flattering. As just one example, we overestimate risks that evoke emotion or are highly salient, and we underestimate risks that fail to give us the chills. As Jason Daley writes in "What You Don't Know Can Kill You," "a risk-perception apparatus permanently tuned for avoiding mountain lions makes it unlikely that we will ever run screaming from a plate of mac 'n' cheese." And with obesity at an all-time high, this presents a pretty substantial problem. But we can use even the bleakest findings to invent new ways to correct for our mistakes and improve our lives. Through such contributions from science, we are able to learn about our flaws and find solutions that mitigate their consequences.

The Downside

The downside is that for each scientific finding, the conclusions we arrive at may limit our options, both professionally and personally. With each science-based recommendation for policies or procedures, we essentially become a more paternalistic society. And most of us are (understandably) uncomfortable with paternalism and its implication of a condescending, one-way relationship in which our lives can be dictated by others. We value freedom for freedom's sake, we value the flexibility to make different decisions depending on our particular circumstances, and we value the ability to choose for ourselves. For all of these reasons, the prospect of setting strict top-down limits is undesirable and complicated.

Nelson Mandela once said, "There is no such thing as part freedom." But was he correct in this sentiment? Or do we value freedom too highly, such that we are harming ourselves as we embark on a relentless pursuit for the freedom to @$#%! up? Alternatively, can we establish a system in which our freedoms are intact but limited in a way that protects us from ourselves? Maybe Robert Frost had a more suitable concept of freedom when he noted, "If society fits you comfortably enough, you call it freedom." Should we relinquish our romanticized vision of freedom and replace it with a more realistic version that accounts for our humanness? Using our scientific toolkit, we can make society comfortable enough to feel free, even if we give some concessions in arenas where it is in our best interest to accept a bit of guidance. For example, we can take advantage of smart feedback systems like those described by Thomas Goetz in "The Feedback Loop" to get people to drive more slowly or take their cholesterol medication. This way, we can simultaneously preserve their freedom to speed, but we can also install "speed meters" that effectively discourage it.

Given the complex costs (limiting freedom) and benefits (better outcomes) that such systems would entail, there is no simple or single answer to how and when we should be paternalistic. Instead, we should consider each case independently and weigh its specific costs and benefits before making decisions that could limit personal freedoms. At the same time, as in all scientific ventures, we should look for principles that can guide our decisions about when to intervene and when to let people make their own mistakes. I would like to propose two such principles: interdependence and agency.

The Principle of Interdependence

John Stuart Mill's influential definition of freedom, which seems to hold up even today, can inform our decisions about when paternalism may be more or less necessary. As Mill said, "The only freedom which deserves the name is that of pursuing our own good in our own way, so long as we do not attempt to deprive others of theirs, or impede their efforts to obtain it. Each is the proper guardian of his own health, whether bodily, or mental or spiritual." In theory, as long as we're not harming others, we can do as we choose. We are free to get blindingly drunk in the comfort of our own homes as long as we harm only ourselves. But as soon as we try to drink and drive, our actions become a serious crime —regardless of the consequences.

With this principle in mind, we can more easily predict some situations in which a paternalistic stance would be highly advisable. For example, one arena where we readily accept a high degree of regulation is driving: we abide by restrictions on age, location, speed, and parking. People are required to wear seat belts, are severely punished if they drink and drive, and risk a hefty ticket when they attempt to use their cell phones. And still, few people advocate for "freedom from seat belts" (although I'm sure there are some libertarians open to this discussion).

In contrast, people are generally opposed to the regulation of private events. If someone chooses to be reckless on his own—juggling chain saws, for example—we consider it up to that person to decide whether it's a good idea or not. I suspect that very few people would want to add an age limit or license requirements to the use of power tools.

However, while the distinction between activities that affect only ourselves and those that affect others appears fairly easy to determine, in many cases it is not—and in a socially connected world, this distinction gets even muddier. For example, is the person who chooses to binge-drink at home, never intending to get behind the wheel, really hurting only himself? He's not endangering drivers, but what if he has a wife and kids? What if he needs a liver transplant and is chosen to receive one over someone who has not put himself at risk for liver damage? I'm not advocating for greater regulation of alcohol sales, but I am suggesting that what can seem like one person's problem may actually leak outside that individual's sphere. As society moves forward, we are beginning to realize how interdependent we are; think about the financial crisis of 2008, when subprime mortgages in the United States brought the world economy to its knees. In areas where our interests and our personal lives overlap and are increasingly intertwined, we may need to adopt more paternalistic methods in order to ensure our safety and well-being.

The Principle of Agency

A second principle that we can use to guide our paternalistic recommendations comes from our view of agency, or the ability of individuals to make good decisions for themselves and for others. The most salient example is that of parenting. We don't exactly trust children to be the best judges of what to eat or whether they need to go to school. If my son were left to his own devices, he would probably build elaborate forts and eat crème brûlée all day. Both David Dobbs's "Beautiful Brains" and David Eagleman's "The Brain on Trial" confront the issue of agency head on; these articles point to the complexities of human development and diversity and how they relate to the decisions we make. And as Eagleman puts forth, "Perhaps not everyone is equally 'free' to make socially appropriate choices." We are not all equipped with the same genes, environmental surroundings, or life experiences (which is, of course, a good thing), and it is unfair to treat everyone in exactly the same way as if we were indistinguishable beings.

But the question of agency is not limited to children and criminals. In fact, much of the research in decision making and behavioral economics highlights how we overestimate the amount of control we have over our decisions and underestimate the forces that influence us. To the extent that we all suffer from these biases and a sense of magnified agency, there must be more cases where a higher degree of paternalism is in order.

From this perspective, let's consider conflicts of interest. While we may realize that such conflicts exist and that they influence *others,* we fail dramatically in perceiving both the extent of their effects and our own susceptibility to them. In one of my favorite studies on conflicts of interest, Ann Harvey, Ulrich Kirk, George Denfield, and Read Montague asked participants to rate how much they liked sixty paintings, ranging from medieval to modern, from the collections of two different art galleries. Half of the participants were told that their payment ($30, $100, or $300) for taking part in the study was sponsored by Gallery A; the other half were told that the sponsor was Gallery B. Each painting had one of the two galleries' logos on it—half of the paintings had Gallery A's logo and half had B's. As you may have predicted, participants tended to prefer the paintings from their sponsoring gallery, which could be explained by a subconscious feeling of the need to reciprocate. However, the story does not end there.

Brain scans of participants showed that the presence of their sponsoring gallery's logo caused increased activity in the parts of the brain related to pleasure (particularly the ventromedial prefrontal cortex, a part of the brain that is responsible for higher-order thinking, including associations and meaning). What's more, this activity increased substantially—along with participants' admiration for their sponsoring gallery's paintings—with each rise in payment. It may seem surprising that a financial favor could influence one's enjoyment of art, especially considering that the favor (payment for participation) was completely unrelated to the paintings. But as with many other types of favors, it was all too easy to create a feeling of gratitude and the desire to reciprocate. And, importantly, none of the participants thought their payment had any influence over their decisions. When asked whether they preferred their sponsoring gallery because of the payment, they scoffed at the mere suggestion.

This lack of appreciation for the power of conflicts of interest is reflected in the real world. A few years ago I gave a lecture on conflicts of interest to about two thousand members of the American Medical Association. I asked how many of the physicians in the audience felt that their judgments were affected by conflicts of interest with their hospital, drug manufacturers, insurance companies, medical device manufacturers, or pharmaceutical sales representatives—and not a single person raised a hand. But when I asked whether they believed that the judgments of other doctors in the room were influenced by conflicts of interest, nearly everyone raised a hand. And again, when I asked whether they believed that the judgments of the *majority* of the MDs in the room were influenced by conflicts of interest, nearly everyone raised a hand.

Once we establish that conflicts of interest create a bias and that we do not appreciate how this bias influences us, we can ask whether we should try to eliminate conflicts of interest altogether. For example, we have laws prohibiting bribery in myriad forms, but what about lobbying? And should we allow companies to finance presidential elections through contributions? Should we allow doctors to receive gifts and benefits from pharmaceutical and device companies? Should we allow bankers to be paid exorbitant amounts when their investments happen to do well and load them first onto the lifeboat when they fail? If we recognize that people are unlikely to see the effects of conflicts of interest on themselves and are therefore unlikely to self-regulate, we should be willing to consider a more paternalistic approach.

Now, for the sake of "freedom of choice," we might decide that we want to allow physicians to consult for medical companies, because without them companies might not improve their products.

We might insist on paying bankers large bonuses even when their investments fail, because otherwise "the smartest and brightest would no longer be attracted to this noble profession." But I suspect that if we truly understood the incredible force that conflicts of interests have in shaping our behaviors, we would not endorse such practices, and we would even attempt to eradicate them.

Where Do We Go from Here?

So here's the issue: if we want to get people to take healthier and better paths in our complex and tempting modern world, we will have to prescribe new instructions, practices, and regulations—and those in turn will inevitably limit our freedoms. Given that most of us are apprehensive about surrendering autonomy, and for good reason, we should not take such decisions lightly. After all, allowing people to make their own decisions is important to our sense of self and independence, and a society without such basic freedoms is unlikely to be successful or happy.

At the same time, we need to acknowledge that our fallibilities and our level of societal interdependence are much more extensive than we give them credit for. With such limited agency and such expansive interdependence, we may want to open the door to science-based paternalism.

How should we determine when and under which conditions we should introduce paternalism or phase it out? In my view, we should apply a careful cost-benefit analysis to each case. First, we should consider whether (or to what extent) a particular case is one in which an individual's actions are likely to damage others. Next, we should determine whether in this case people are likely to naturally make decisions that are in their best long-term interest. If the answers to these two questions are "yes" and "no," respectively, we should empirically examine a range of potential interventions—starting with small-scale studies—to measure the potential costs and benefits of various solutions. To retain personal freedoms, we should begin by experimenting with more hands-off approaches that provide better information (such as posting calorie information in restaurants and measuring whether this has any effect on the consumption of unhealthy foods). And only if we can document empirically that these interventions fail or fall short of expectations should we move to more invasive and paternalistic interventions. For example, in the case of posting calorie information, the evidence across multiple studies (for example, see the 2012 study by J. A. Schwartz, J. Riis, B. D. Elbel, and D. Ariely) is quite clear that merely providing calorie information does not encourage healthier eating. This means that it is time to try more aggressive approaches, such as taxation or smaller default portion sizes.

In the end, the level of paternalism we want in our society is not going to be fully determined by science because there is no way to fully capture the costs involved in restricting the freedom of people and societies. Perhaps one day we will figure out how to measure such costs, but until then we should use science as an input to help us understand which areas of life we should regulate to a higher degree and to come up with interventions that balance effectiveness with minimum impact on personal freedoms. It is not going to be easy, but as our living environments become more and more complex, with more and more temptations surrounding us, it is going to be even more important to figure out where we can succeed on our own, when we need a little helping hand, and where we need the strength of a whole arm.

For me, this is one of the main goals for science in the years to come—to figure out the human condition and design our environment to reduce our tendency for error and maximize our potential.

Irrationally Yours,
DAN ARIELY

P.S. My deepest thanks go to Aline Grüneisen, in particular for help with this book.

References

Harvey, Ann, et al., "Monetary Favors and Their Influence on Neural Responses and Revealed Preference," *Journal of Neuroscience* (2010).

Schwartz, J. A., et al., "Inviting Consumers to Downsize Fast-Food Portions Significantly Reduces Calorie Consumption," *Health Affairs* 31, no. 2 (2012): 399–407.

PART ONE

Bacteria/Microorganisms

BRENDAN BUHLER

The Teeming Metropolis of You

FROM *California Magazine*

YOU ARE MOSTLY not you. That is to say, 90 percent of the cells residing in your body are not human cells; they are microbes. Viewed from the perspective of most of its inhabitants, your body is not so much the temple and vessel of the human soul as it is a complex ambulatory feeding mechanism for a methane reactor in your small intestine.

This is the kind of information microbiologists like to share at dinner parties, and you should too, especially if you can punctuate it with a belch.

It's not that our bodies aren't enormously interesting in themselves, only that they are not the whole story—not nearly. Each of us is a colonized country, a host of multitudes, a reservoir of biodiversity. Indeed, the species of bacteria living on your left hand are different from those living on your right.

At the dawn of the twenty-first century, navel gazing is on the frontiers of biology and medicine, as proven by the Belly Button Biodiversity Project at North Carolina State University. They take swabs, sequence DNA, and post photographs of the resulting cultures. (Request a kit online!)

And our biota—all the living creatures in and on us—is not a set of merely passive passengers. The bacteria in our gut aid our digestion and, as we are increasingly discovering, help defend us from pathogens. There are a couple of ways of thinking about how they protect us, says Russell Vance, a University of California, Berkeley, professor studying the interactions of bacteria and the immune system. The first protection is by simply existing.

Think of your body as a big city apartment building. Our normal biota comprises its tenants, and they're solid folk. They keep up the maintenance, take out the trash, and pay their rent—that is to say, they promote healthy tissue growth, comprise the majority of the dry mass in our feces, and, by fermenting carbohydrates, provide us with roughly a quarter of our calories. Just by maintaining building occupancy, they keep bad elements from moving in as squatters, beating up the superintendent, ripping out the copper pipes, and turning the whole place into a crack den—that is, they compete for nutrients, occupy the mucous lining, and sites in the intestine where pathogens might attach and attack.

Another way our tenants protect the place is by actively policing the hallways—a healthy biota makes the pH of our guts inhospitable and even toxic to many pathogens.

What's more, the bacteria in our guts act as a scrimmage team for our immune system. Over our lifetimes, our bacteria and our immune systems compete with each other, but for reasons scientists are still figuring out, never too aggressively. They take the field against each other, but they don't play tackle football.

To some extent, this makes sense; after all, if harmless bacteria were playing for keeps, they wouldn't be harmless—they would be attacking our intestinal cells and colonizing the rest of our body. Similarly, if our immune systems were fighting at full strength, we would be sick all the time, since most of the experience of being ill—fevers, aches, coughing, runny noses, and so on—is the result of our body trying to make itself inhospitable to microorganisms.

Compelling evidence for our biota's role in our immune system comes from the laboratory's quick-breeding little helpers: mice. It's possible, Vance says, to breed sterile—as in germ-free—mice, ones with absolutely no gut biota. These mice live in sterile cages, eat special food, and breathe filtered air. If they are taken out of their cages, their immune systems collapse when exposed to the meekest pathogens, and the mice die.

Unfortunately, there are cases when the body's biota and its immune system stop their friendly scrimmage and start playing like the 1970s Oakland Raiders. This is thought to cause or play a role in autoimmune illnesses, including environmental allergies, Crohn's disease, and inflammatory bowel disease.

Vance's lab at Cal uses mice to study how our body's innate immune system distinguishes between pathogens and our normal gut bacteria. So far, it seems that our body is guarded not so much by watchdogs as by burglar alarms: they can't differentiate friend from foe. All they can do is make themselves known if something gets in.

Vance is looking at one particular burglar alarm, a protein in the cell that under certain circumstances will signal the immune system to attack all bacteria, including healthy gut bacteria. It's not clear exactly when those circumstances might occur, but the fact that this protein doesn't distinguish between good and bad bacteria could help us understand what triggers autoimmune diseases.

Life, however, is struggle, and our immune system and friendly gut bacteria are not the only players in the game. Sometimes a pathogen causes our immune systems and our biota to turn against each other. Research by Andreas J. Bäumler at the University of California, Davis, shows that *Salmonella typhimurium,* a bacterium that causes gastroenteritis, actually benefits from our immediate immune response. This reaction is a flood of antimicrobial agents and an overall change in your intestinal environment that is very noticeable to you and the person in the bathroom stall next to you. This changed environment is hostile to all bacteria, including your own. In fact, it is more harmful to your biota than it is to salmonella. Because your immune system is better at thrashing your biota than it is at thrashing salmonella, the salmonella suddenly has more nutrients to eat and more space to grow.

We are just beginning to understand the role our biota plays in human health and disease, says Rob Knight, a biophysics researcher at the University of Colorado, Boulder. Knight works on the Human Microbiome Project, sequencing and analyzing the genetic makeup of our biota, work that has only recently begun to take off as gene sequencing has become much cheaper and faster.

Right now, he says, we're still not sure how many species we play host to. In our gut alone there may be anywhere from one hundred to a thousand species. (The answer depends not only on how many we can discover but on how finely one defines "species.") And the particular makeup of the community living in one's gut varies wildly from individual to individual.

It seems we acquire our first colonists literally at the moment of birth. You get most of it from your mother, but even identical twins don't end up with identical biota, and a sibling born by cesarean section will have a very different set than his vaginally birthed brother—the C-section child is exposed to the biota on his mother's skin, while the child passing through his mother's vagina meets biota more like those in mom's guts. The differences between unrelated strangers are even more pronounced, Knight says. Two unrelated North Americans will share only 10 percent of their intestinal bacteria, and a North American and a South American will share only 5 percent.

And the differences in our biota have a wide-ranging effect on our health. For one, it could even be that our biota is making us fat.

Knight points to a series of experiments in which the gut biota is transplanted out of obese mice and into healthy ones, whereupon the healthy mice either become obese or develop colitis. This is why Knight is interested in the potential of healthy bacteria called probiotics to treat human obesity and other conditions. "We're probably where, I don't know, radium was in 1910, where people thought making everything radioactive would just be great," Knight says. "And while that wasn't true, it doesn't proscribe the important uses of radiation in, say, cancer therapy."

So far probiotics are more popular as a marketing term for nutritional supplements and bacteria-enriched yogurt than as recognized medical treatments. In the coming years, Knight says, probiotics will likely become an increasingly legitimate medical tool. But not quite yet. There have been some very promising studies where mice were successfully treated with probiotics for inflammatory bowel disease. "We're really good at curing diseases in mice and somewhat less good at curing them in humans."

Scientists are only now beginning to figure out just how important our biota can be. For instance, Knight says, one study has shown that these harmless bacteria can change our metabolism in a way that affects our behavior, and not only behavior related to eating.

Again, mice are involved. Take some lab-grown sterile mice. Take some normal bacteria-infested mice. Run them both through a maze that includes a high beam without rails. It turns out that the sterile mice are much bolder and will spend more time on the beam than their normal cousins.

Even more interesting, if you introduce normal bacteria into adult sterile mice, their behavior doesn't change. If, however, you take mice that are still growing and colonize them with the same bacteria, they will develop more typical fear responses. This suggests that our biota can influence brain development. It makes a certain amount of sense, Knight says, considering that our guts produce many of our neurotransmitters, including 90 percent of our serotonin, a linkage that, in a term of complete scientific awesomeness, is called the "gut-brain axis."

In other words, it's looking increasingly likely that it's not so much that you are mostly not you as that you are also the slime sloshing around your innards. If the medical advances of the twentieth century taught us to stop thinking of our minds and bodies as separate entities, the twenty-first may teach us to include our symbiotic biota: Our Bacteria, Ourselves.

VIRGINIA HUGHES

Our Body the Ecosystem

FROM *Popular Science*

WHEN JAKE HARVEY VISITS the clinical center at the National Institutes of Health in Bethesda, Maryland, he is usually dirty, itchy, and wheezing—not the happiest state of affairs for a fourteen-year-old boy. But his doctors require that for twenty-four hours prior to each visit, he refrain from bathing, using the inhaler that soothes his asthma, or applying the ointment that softens his eczema. In order to study his illness, they need him to be as close to his natural state as possible.

Jake's discomfort could lead to better treatments for the millions who have eczema—a disorder marked by dry red rashes in the creases of elbows, behind knees, and on the back of necks—as well as an array of other allergic reactions. By understanding eczema in a new way, as the product of a delicate interaction between the immune system and the legion of bacteria that live on the skin, one group of scientists hopes to better understand what triggers it and why the number of diagnosed eczema cases in developed countries has dramatically increased over the past few decades.

These researchers, led by Heidi Kong, a dermatologist at the Center for Cancer Research at the National Cancer Institute, and Julie Segre, a geneticist at the NIH, are just one part of the five-year, $173 million Human Microbiome Project (HMP), an effort to characterize the thousands of species of microbes that live on or in us. So far, Jake has made half a dozen trips to Bethesda, sixty miles each way, to donate a few skin cells to the project.

Jake has been struggling with eczema since he was a few months old. The rash never stops itching, and when he scratches, it bleeds and scabs and gets even itchier. His clothes stick to the sores. He has tried many treatments, including petroleum jelly, topical steroids, antibiotics, and also dairy-free, gluten-free, and probiotic diets. None of them has worked very well. When he was younger, he went to school with bandages on the tips of his fingers and slept with socks over his hands. In bed, he still sometimes lies on his back with both legs sticking straight up so they're easier to scratch. "I've never really gotten a full night's sleep," he says.

As many as 30 percent of all children develop eczema, and no one knows what mix of genetic and environmental factors sets it off. The disease runs in families, yet Jake's twin sister, Becca, has perfect skin. For about 60 percent of children with the disease, it goes away by early adolescence. The others frequently deal with outbreaks for life.

Whether the rash disappears or not, nearly one third of children with eczema go on, like Jake, to develop asthma and hay fever. Asthma and hay fever also involve inflammation, but very little is known about what quirk of the immune system links them all together. "We've been to pediatricians, allergists, dermatologists for years," says Jake's mother, Debbie, "and nobody can figure him out."

The average human body is made up of trillions of cells. The average human body also houses about *10 times* that number of bacterial cells. Scientists have been curious about our bacterial cohabitants since 1683, when Anton van Leeuwenhoek, using a microscope he had built himself, examined his own dental plaque only to discover "little living animalcules, very prettily a-moving." But it has only been within the past few decades that scientists have begun to understand just how many varieties of bacteria live in or on our bodies. And now they increasingly suspect that many diseases are caused not by individual bacteria but by the delicate interplay between multiple bacterial species and the human host.

In the Human Microbiome Project, researchers plan to characterize the vast numbers of bacteria, fungi, protozoa, and viruses in our body by sequencing their genes. That won't be easy. In the past, researchers had to grow each species outside the body before they could identify it, a process that required intense research to determine optimum growing conditions. Only the hardiest and most numerous bacterial species—for example, *Staphylococcus aureus* and *Streptococcus pyogenes,* which can cause life-threatening infections—have been thoroughly studied in the laboratory. Now advances in DNA sequencing—the very same that made it possible for the Human Genome Project to decode the 3 billion base pairs of our own genome quickly—have provided the technology to make a comprehensive Human Microbiome Project possible.

"What we want to understand first," says HMP coordinator Lita Proctor, "is what's considered the norm. What does a typical healthy human have?" HMP researchers are building a reference database of the genetic fingerprints of about three thousand different bacterial species. Scientists are also thoroughly characterizing the makeup of microbial communities found at half a dozen body sites—including the gut, the mouth, the skin, and the groin—of three hundred normal people. The next step is to compare those results with what researchers find in patients with specific medical conditions, such as eczema, Crohn's disease, and ulcerative colitis.

The skin samples from Jake and other children with eczema will help Segre and Kong determine whether changing profiles of skin flora, and their interaction with the human immune system, are involved in the rising rates of the disease. Some 34.1 million Americans suffer from asthma, and up to 50 million have seasonal allergies.

"In the last three decades, all of these allergic disorders—asthma, eczema, hay fever—they've all tripled," Segre says. In that short a time frame, the culprit can't be simply changes in our own genome. "So it must be something about the gene-environment interaction. And I now believe that that's modulated by the body's bacteria."

The first part of the exam is like any other doctor's visit, Jake says. Sitting on a stool next to him, Kong asks a series of questions. *Which medications are you currently taking? Does your school use antibacterial soaps? How bad are your allergies this month?* On a diagram of a body, she notes all the spots where Jake's rash has cropped up that day. Then she pulls on a pair of blue examination gloves and turns to a tray containing a row of sterile swabs and scalpels. The swabbing comes first. Jake stretches out his right arm, palm up, and Kong firmly rubs a wet foam swab in a circular motion in the crook of his elbow; the sterile water solution stings when it hits his open sores. Kong drops the tip of the swab into a plastic tube filled with the same liquid and places the tube in a bucket of dry ice. Then comes the scalpel. Kong moistens the blade and gently scrapes some skin from a spot next to the area she swabbed. She wipes the flakes from the blade with another swab and drops the tip into another plastic tube. Kong repeats the procedure on each elbow, each inner forearm, the back of each knee, and finally in one of his nostrils. Once the sampling is over, she and her colleagues take pictures of Jake's rash. Then they check his vital signs.

Jake doesn't know much about what happens to his skin cells once they disappear into the lab. For the most part, he comes to the clinic, which he found because he was enrolled in an NIH study on asthma, for Kong's helpful advice about his eczema. (She recommended that he take diluted bleach baths during bad flares, which has given him some relief.) But he says he's proud to be part of the effort to understand this disease, in case his future children end up with the same troubles.

Jake's skin samples, like those from the more than twenty other children participating in the study, wind up in a nearby laboratory complex, where they are stored in a large freezer set at –112 degrees F. The inside of the door is lined with thick white snow. Clay Deming, a biologist at the lab, explained to me how he prepares each sample. He selects a tube and defrosts it; adds lysozyme, an enzyme that breaks down bacterial cell walls; and shakes the concoction using a vortex mixer. This releases DNA from the bacterial cells so that it floats freely in the liquid. He then pours the solution over a gel filter designed so that just bacterial DNA fragments will stick to it.

With the DNA fragments thus separated, Deming can use a technique called polymerase chain reaction to make millions of copies of one particular bacterial gene: 16S. All bacteria carry the gene, but its DNA sequence varies from species to species. After amplifying the DNA fragments, Deming is left with a new set of tubes, each holding the 16S pieces harvested from a particular area of the participant's skin.

He puts the tubes on ice in order to FedEx them to a building five miles north, the NIH Intramural Sequencing Center. There, machines read the precise sequence of 1,500-odd letters in each 16S fragment, giving the chemical signature of each species. Finally, technicians upload the sequence data to an internal network, ready to be mined by Segre and Kong.

The skin is an ecosystem. Like any other ecosystem, it harbors permanent residents and also migrant species that flock to a few hot spots during certain seasons. Those fluctuations powerfully influence how the skin works. *Staphylococcus epidermidis,* for example, may help educate the skin's immune system, training it to recognize particular molecules so that it can better respond to an attack by harmful species. *S. epidermidis* churns out proteins that prevent unwanted invaders from adhering to skin. So it makes sense that disrupting these complex microbial interactions could lead to skin problems.

With the eczema study, Segre says she hopes to find patterns in the microbiome that could predict the onset of an eczema flare in a particular child or even help doctors choose a far more effective treatment; for some patients, bleach baths might help, whereas for others, a round of anti-inflammatory steroids might be the best choice.

Kong and Segre are not the only scientists looking into the connection between the microbiome and inflammatory skin conditions. Martin Blaser, a microbiologist and physician at New York University, thinks the microbiome also plays a role in psoriasis. Some twenty-five years ago, Blaser's father had told him that his psoriasis—scaly red-and-white patches on the skin—seemed to get better after he started taking allopurinol, a medicine that treats gout. Blaser, who has mild psoriasis himself, took notice. He knew that because allopurinol interferes with the synthesis of DNA, it kills bacteria or makes it very difficult for those bacteria to grow.

His father's experience spurred Blaser to look deeper into the connection between allopurinol and microbes. He discovered that a few decades earlier, researchers had run clinical trials on allopurinol as a treatment for psoriasis. The results were inconclusive, but Blaser thought that the medicine might work on a particular subgroup of psoriasis patients.

In 2002, he finally secured funding to study the microbiomes of psoriasis patients. For the project, Blaser performed a general census of bacteria living on the inner forearms of six healthy people. Later his group did the same analysis on skin from six people with psoriasis, and they found some striking differences. The six psoriasis patients had considerably more bacteria from the Firmicutes phylum than people without psoriasis. What's more, the most common class of bacteria in healthy skin, Actinobacteria, was significantly underrepresented in the lesions.

"Those results, which were important at the time, now are kind of laughable because they're on such a small scale," Blaser says. Now, with funding from the Human Microbiome Project, his team is looking in much greater detail at how the overall microbial numbers, as well as changes in the species distribution, vary in patients with psoriasis.

In May 2009, Kong, Segre, and their colleagues published the first comprehensive catalog of skin microbes, based on samples taken from twenty different body parts—from the oily crease outside the nose to the moist spots between the fingers—of ten healthy people. By sequencing hundreds of the 16S genes in each sample, the researchers found that our epidermal ecosystem is much more diverse than anybody had thought, with discrete bacterial populations ranging in size from fifteen species behind the ear to forty-four on the forearm. The team also looked at whether the microbiome is consistent across individuals; that is, do some individuals always have a certain set of bacterial species, no matter what part of the body those bacteria live in? The researchers learned that, no, the bacteria are finely tailored to specific locations on the body rather than to individual humans. For example, the bacterial communities under your arm are more similar to those under someone else's arm than to those behind your knee.

Jake is allergic to many things—mold, peanut butter, dogs, cats —and his asthma is severe. In 2009 he almost died because he couldn't breathe. "The scary thing was, within seconds, he was in respiratory and cardiac arrest," his mother recalls. "No warning. No lip swelling or dizziness or mouth burning. He just went down."

Eczema's strong link to allergies suggests that the immune system is key to the development of the disease, and researchers have focused on inflammation for decades. So how does the skin microbiome fit into this picture? In 2006, a coalition of researchers from the United States, France, Ireland, and the UK discovered that up to half of people with eczema have mutations in filaggrin, a protein in the skin barrier, the top layers of the skin. Now some researchers suspect that this flaw in the skin barrier allows entry to particles that trigger the immune response and creates an ecological niche for a different set of microbes.

To uncover any link between the skin barrier and microbes, Kong and Segre are sampling Jake and other children with eczema at three points: during a normal, or baseline, period; during a flare; and two weeks after treatment. In a preliminary analysis of data from ten patients, Kong and Segre have confirmed older studies showing that there are huge amounts of *Staphylococcus aureus* on the skin during a flare. What wasn't known before, however, is that *S. aureus* crowds out the other bacterial species during a flare. Kong and Segre are trying to figure out how treating these patients changes their bacterial diversity and leads to better individual results.

Segre doesn't know whether bacteria associated with eczema are the cause of the disease or simply a consequence of living with it. To find out, she plans to perform a metagenomic analysis of the samples. During a metagenomic analysis, scientists compare thousands of genes present in a particular species' DNA. By looking at the biological function of the genes—what kinds of proteins they make and what kinds of biological pathways those proteins are involved in—the scientists can make educated guesses about the role of each species and how different species may work with one another and with our own genome.

"We all believe there's an interdependency among these organisms. They're highly dependent on their neighbors for their survival," says Claire Fraser-Liggett, director of the Institute for Genome Sciences at the University of Maryland. Metagenomics, however, is immensely complicated. Researchers know little about how the millions of microbial genes might work together, and it's difficult to sort out which patterns are signatures of disease versus part of normal variation between people. "There's no way to overemphasize the analytical challenges," Fraser-Liggett says. "It's something that everybody is struggling with."

Early results from Segre's study indicate that researchers might not have to decode the entire microbiome to better treat children's skin diseases. She says she envisions a day in the not-too-distant future when a dermatologist, during an office visit, will drop a skin sample into a machine that spits out a signature of the microbiome.

If she finds certain microbial profiles that predict the onset of an eczema flare, for example, doctors could use that data as a guide for action. They might tell the patient to take a few extra bleach baths that week or to skip football practice. Researchers might even be able to create "probiotic" concoctions to replace the bacterial species that patients lack during a flare, Segre says. Hospitals are already routinely swabbing people's noses to screen for drug-resistant bacteria. "You get that result in less than an hour," Segre says. If the screen turns up MRSA (methicillin-resistant *S. aureus*), for example, doctors can prescribe an antibiotic known to be effective against the species.

These potential applications are many years off, and Segre's initial studies probably won't have much effect on Jake's eczema. Still, the Harveys are happy to be moving the field forward. "It's probably going to be a while before Jake's helped," Debbie says. "But in the future, if someone else can avoid sitting up all night scratching their legs, that would be great."

JEROME GROOPMAN

The Peanut Puzzle

FROM *The New Yorker*

JILL MINDLIN PRIDES HERSELF on being a good parent. An attorney who lives on the North Shore of Long Island, she read books about how to raise healthy and happy children and dutifully followed their advice. She bought the car seat with the highest safety rating and covered her son and daughter with sunblock whenever they went outside. With her pediatrician's approval, she breast-fed her children until they were at least a year old and gave them "no formula whatsoever" and no milk products or peanuts. As the American Academy of Pediatrics recommended in 2000, she introduced solid foods slowly and in small amounts.

In 2002, when her daughter, Maya Konoff, was nine months old, Mindlin took Maya for a checkup, and she got several immunizations. After they came home, Mindlin gave her a little yogurt. Soon, Mindlin told me, "Maya blew up like a tomato, bright red, swelling from head to toe." She called the office, assuming that her daughter was reacting to the immunizations. The pediatrician told her that it was more likely an allergy of some kind. "Fortunately, there was liquid Benadryl in the house, and I was able to get Maya to take some," Mindlin said. The reaction slowly subsided.

Several days later, Mindlin took Maya to see a pediatric allergist at a hospital on Long Island, and he told her it was unlikely that her daughter had a dairy allergy, since she had been breast-fed and was on a restricted diet. But Mindlin asked that Maya be examined, and the allergist placed a small amount of milk protein under the baby's skin. Within minutes, she broke out in hives. As it turned out, Maya was also allergic to eggs, peanuts, tree nuts, and sesame seeds.

Despite her mother's vigilance, Maya has had other frightening reactions. On a family outing to the Long Island Children's Museum a few months later, after eating something labeled "vegetarian cheese," Maya struggled to breathe and then lost consciousness. On vacation in South Carolina in 2003, Maya wanted a hot dog. "We asked the waiter to be sure that there were no dairy products in the food," Mindlin recalled. "He came back to the table and said that the package said a hundred percent beef." But a few minutes after eating the hot dog Maya began vomiting and swelling up. Mindlin later learned that the hot dog contained a milk protein. This time the doctor in the ER gave Maya an epinephrine injection. Epinephrine, another term for adrenaline, can rapidly shut off a severe allergic reaction, and Mindlin now makes sure there are syringes of it in each of her handbags and in Maya's knapsack.

Dr. Hugh Sampson, the director of the Jaffe Food Allergy Institute at Mount Sinai Medical Center in New York and an international expert on food allergy, is Maya's doctor. He is a tall sixty-year-old with an athletic build and a full head of graying hair. Sampson and Dr. Scott Sicherer, a pediatric allergist who is also at Mount Sinai, have conducted extensive studies throughout the United States that show that the rate of allergy is rising sharply. Sampson estimates that three to five percent of the population is allergic to milk, eggs, peanuts, tree nuts, or seafood. In the past decade, allergies to peanuts have doubled. Other researchers have found the same phenomenon in Great Britain. "This increase in the incidence of food allergy is real," Sampson said when we spoke recently. He cannot say what is causing the increase, but he now thinks the conventional approach to preventing food allergies is misconceived. For most of his career, he believed, like most allergists, that children are far less likely to become allergic to problematic foods if they are not exposed to them as infants. But now Sampson and other specialists believe that early exposure may actually help prevent food allergies.

Sampson recalls that in 1980, when he started researching the subject as a fellow in immunology at Duke University, "food allergy was not a field that anybody wanted to get into." Many doctors said that patients who claimed that food allergies were causing stomachaches and rashes were often just manifesting psychosomatic symptoms. "I approached the subject with the assumption that I would prove it didn't exist," Sampson said.

In one early test, he gave a girl in the first grade a bit of egg camouflaged in applesauce. To Sampson's astonishment, she started wheezing and projectile vomiting. Five years later, he found that his one-year-old daughter was allergic to eggs. As Sampson got deeper into his work, he was struck by how little was known about the condition. No one knew why some children react to a food protein when it is placed on their skin but not when they eat it or why others have antibodies in their blood that predict allergic reactions they don't end up having.

Sampson watched as the incidence of food allergies rose alarmingly in the West while cases remained rare in Africa and Asia. He and other researchers began to investigate whether the problem could be prevented if Western mothers continued breast-feeding as long as possible. This would keep their babies away from potentially allergenic foods until their immune systems had developed sufficiently. Laboratory studies reinforced the theory. Sampson's research group and others found that mice that had never been exposed to a particular food protein couldn't mount an allergic reaction to it. This suggested that isolating young children from even minor exposure to potentially allergic foods would be beneficial.

In 1989 Dr. Robert Zeiger, a pediatric allergist and immunologist at Kaiser Permanente Medical Center in San Diego, published related results from one of the only controlled research studies on the subject. In the Zeiger study, which appeared in the *Journal of Allergy and Clinical Immunology*, mothers prone to allergy were randomly assigned a restricted diet. They avoided cow's milk, eggs, and peanuts during the last trimester of pregnancy and during breast-feeding; their infants were given the supplement Nutramigen, derived from casein, and kept off all solid foods for six months; cow's milk, corn, soy, citrus, and wheat were prohibited for twelve months, and egg, peanut, and fish for twenty-four months. After one year, the infants on the restricted diet had significantly fewer allergies than those in the control group. "Reduced exposure of infants to allergenic foods appeared to reduce food sensitization and allergy primarily during the first year of life," Zeiger wrote.

A few experts believed that Zeiger's research had not yielded results from which one could draw major conclusions. But Sampson was influenced by the article, and most of the other leading thinkers in the field agreed with the findings. "We know that the human immune system is immature for the first year or so. So I was thinking initially that as long as we don't expose babies to a food, they can't make an immune response," Sampson said, "and if we can wait until their immune system matures after a few years, they could do better when later exposed to the food."

In 1998 the Department of Health in the United Kingdom issued guidelines for doctors and families codifying these recommendations. In 2000, the American Academy of Pediatrics did the same.

The proteins in eggs, milk, peanuts, tree nuts, fish, shellfish, wheat, and soy that trigger allergic reactions don't readily decompose when exposed to heat in certain types of cooking or to the acid in our stomachs. Within the gastrointestinal tract, the immune system battles pathogens while it ignores harmless food proteins and allows nonthreatening bacteria to reproduce. Proteins that are easily broken down by heat or digestion, such as many of those found in fruits, generally pass by. Proteins that resist breakdown are more likely to stimulate an allergic reaction.

People with the worst food allergies usually have very high levels of an antibody called immunoglobulin E (IgE). When someone like Maya drinks milk, the IgE grabs hold of specific proteins that trigger the body's release of potent molecules like histamine and cytokines. The immune system overreacts to fight the protein that most people's bodies ignore. When Maya "blew up like a tomato" and stopped breathing, it was because these molecules created so much swelling and inflammation that her throat closed up. For reasons that are still not completely understood, some people manifest their allergic reactions with nothing more than an outbreak of eczema. While there is a genetic predisposition to food allergies, no one has identified the specific genes, and there is no biological explanation for their existence.

"From an evolutionary biology point of view, food allergy makes no sense at all," Dr. Scott Sicherer, Sampson's colleague at Mount Sinai, said. Hunters and gatherers who had potentially fatal reactions to tree nuts, peanuts, seeds, and fish would be at a distinct evolutionary disadvantage and were less likely to pass on their DNA to progeny. "It seems pretty clear that food allergy is a condition that resulted from the environment we created," Sicherer said.

One explanation for the rise in food allergies is called the "hygiene hypothesis." The natural environment exposes us to microbes that help teach our immune system to differentiate between dangerous pathogens and nonthreatening nutrients. When we shield children from dirt in the playground and from sick kids in preschool, we may limit their infections while also reducing their exposure to healthy microbes. This could make them susceptible to food allergies. Studies of mice raised in a germ-free environment show that they have abnormal immune systems and are more prone to allergic reactions. It is possible that we are doing the same thing to ourselves.

Researchers have also proposed several theories based on observations of geography and diet. Vitamin D is believed to reduce the development of allergies, and sunshine promotes vitamin-D production. Doctors in cold parts of the United States write three or four times as many prescriptions for epinephrine to treat food allergies as do doctors in warm locales. Dietary changes might also play a role. Eating more animal fat can increase the presence of a chemical, prostaglandin, that contributes to the body's inflammatory responses. And as people also eat fewer fresh fruits and vegetables, they fail to take in substances, such as beta carotene, that limit inflammation in tissues.

One of the few pediatric allergists who questioned the guidelines written in 1998 and 2000 was Dr. Gideon Lack, at St. Mary's Hospital in London. Lack studied philosophy and psychology before medicine, and his background is evident in his approach to science. "If eating eggs or eating peanuts in an allergic sufferer causes a reaction, then clearly the way to prevent a reaction from occurring is by not eating egg or peanut," he said. "That makes sense. But that's different from saying that clearly the way to not become allergic in the first place is not to eat egg or peanut."

Lack published letters in *The Lancet* and the *British Medical Jour-*

nal that pointed out the absence of compelling evidence used to support the expert guidelines. His skepticism was not well received. "It was very hard to get any grant support to study my ideas," he said.

In 2003, Lack gave a lecture in Israel about the apparent rise of peanut allergies in the United Kingdom. "It was a large lecture hall in Tel Aviv, filled with pediatricians and allergists. And I asked them, 'How many of you have seen a case of peanut allergy in the past year?' Something like three hands shot up." Lack told me that if he had asked that question in the United Kingdom, 90 to 95 percent would have raised their hands.

Working with researchers in Israel, Lack surveyed more than five thousand children in Jewish schools in North London and more than five thousand schoolchildren in an ethnically and economically similar region of Tel Aviv. The team obtained detailed information about the families' consumption of foods like peanuts, sesame, and tree nuts. They also cataloged other allergic diseases, such as asthma, eczema, and hay fever. The risk for peanut allergy among Jewish children in the London area was nearly eleven times higher than among those in Tel Aviv. Tree-nut allergy was fourteen times higher, and sesame five times higher in the United Kingdom. The relative risk for milk and egg allergy was about two to three times higher.

Lack's study does not offer any proof about the cause of the variance in allergies between Jewish children in London and in Tel Aviv, but he believes the striking discrepancy may be due to a difference in diet between Israel and England. "The joke in Israel is that the first three words a child says are *abba,* meaning 'father,' *ima,* meaning 'mother,' and Bamba," Lack said. Bamba is a peanut concoction that looks like a Cheez Doodle, and it is a staple of infants' diets in Israel.

Lack did part of his training in pediatric allergy at the National Jewish Medical and Research Center in Denver, where he discovered that mice could develop allergies to a particular egg protein that was first rubbed on their skin or inhaled before they had ever eaten it. He wondered whether children in the United States and the United Kingdom might become allergic to peanuts through a similar mechanism. In a study published in the *New England Journal of Medicine* in 2003, he reported that children with eczema had often been previously exposed to an ointment containing peanut oil and were later found to be allergic to peanuts. He also determined that there was no correlation between women who had eaten peanuts while pregnant and the development of peanut allergies in their children. His study challenged the idea that restricting a mother's diet would prevent peanut allergy and highlighted how children can inadvertently be exposed to food proteins.

In 2006 Lack received support from the National Institutes of Health as well as from two charitable organizations, the Food Allergy Initiative and the Food Allergy and Anaphylaxis Network. He is now more than halfway through the LEAP study—Learning Early About Peanut Allergy. Six hundred and forty babies have been enrolled in the trial. The children are randomly selected either to eat peanut products or to avoid them entirely. The study will compare the rates of peanut allergy between the two groups. Lack is also conducting a study funded by the Food Standards Agency and the Medical Research Council in the United Kingdom about when to wean children from breast-feeding and how a baby's consumption of allergenic foods affects her later development of allergies. As part of that work, he is examining thirteen hundred babies in the United Kingdom.

Lack believes that a child becomes tolerant of a variety of food proteins through exposure in the first six months of life. In developing countries, he notes, children often consume solids, initially chewed by their parents, at two or three months. "Years ago, nobody had blenders or food mixers, and today in developing countries people still don't. The easiest way to get solid foods into a baby's mouth is to chew it up, so it's moist and coated with saliva, and then spit it into the baby's mouth."

A paper published in *Maternal and Child Nutrition* in January 2010 reported that some two-thirds of students at a university in China were given premasticated food as infants. Only about 14 percent of American infants receive solid foods in this way. Saliva is a rich source of enzymes that can help break down solid foods and of antibodies that might coat food proteins in a way that makes them less allergenic to infants.

Lack's research has gradually gained influence with leading allergists, including Hugh Sampson. By 2006 Sampson realized that his recommendations about food avoidance did not conform to what he termed "the real world." Doing nothing more than inhaling or touching an allergen could prompt a reaction in some children. "You can't avoid food proteins," Sampson said. "So when we put out these recommendations we allowed the infants to get intermittent and low-dose exposure, especially on the skin, which actually may have made them even more sensitive."

Sampson believes that some 80 percent of infants who are allergic to eggs or milk will outgrow the allergy by their teenage years and that preventing them from being fed products with these foods may prolong the time that takes. "I spent most of my career telling mothers to avoid these types of foods for their babies," he told me. "Now we're testing to see if we should advise mothers to give the foods to them."

In January 2008, the American Academy of Pediatrics released a clinical report by Mount Sinai's Dr. Sicherer and other researchers that overturned the expert advice of the past decade: "Current evidence does not support a major role for maternal dietary restrictions during pregnancy or lactation. . . . There is also little evidence that delaying the timing of the introduction of complementary foods beyond four to six months of age prevents the occurrence of [allergies]." Dr. Frank Greer, a specialist in newborn nutrition at the University of Wisconsin School of Medicine and Public Health and an author of the clinical report, told me, "There is so much out there about how to feed infants, when to begin rice cereal, how to phase in yellow vegetables and then green vegetables, that has no basis in scientific evidence. It's not surprising that recommendations were made which were based on so little data."

Dr. Susan Baker, a professor of pediatrics at the State University of New York at Buffalo and an expert on nutrition for children, chaired the committee overseen by the AAP that released the recommendations in 2000. She told me that safety concerns drove the experts to recommend restricting exposure of infants to potentially allergenic foods, particularly cow's milk. "At the time, there was a proliferation of infant formulas on the market. Babies not only have cow's milk allergy with eczema, but some who are intolerant of milk also develop bloody diarrhea. The real concern was that the formulas might do harm. That sort of propelled us." The committee, she said, moved from milk products to restricting other allergenic foods, like peanuts and fish. "We in medicine are making a lot of decisions and recommendations based on not a lot of solid evidence. So you toe a fine line. You want to try to get pediatricians something that is as good as it can be to help guide their practice and their thinking. Did we overreach with peanuts and other foods? Probably. Could it have been better? Absolutely."

The 2000 recommendations have been overturned, but Gideon Lack is disturbed by what families now face. "Basically, we are all in limbo," he said. Sicherer told me, "This is a tricky area. The AAP has backed away from making recommendations, since the evidence is weak. I try to emphasize with my patients not to feel guilty that they did or did not do something that would have resulted in their child having a food allergy. Even the experts are not certain what to advise."

People with food allergies live under a constant threat in a society that is still poorly informed about the condition. For people with peanut and tree-nut allergies, incidents in restaurants account for nearly a quarter of unintentional exposures and about half of all fatal reactions.

In 2007 Sicherer published the results of a survey of a hundred managers, servers, and chefs in establishments ranging from continental restaurants to bakeries and delis. Focusing on New York City and Long Island, Sicherer found that about a quarter of managers and workers believed that consuming a small amount of the allergen would be safe; 35 percent believed that frying would destroy it; and a quarter thought it was safe to remove an allergen from a finished meal, like taking walnuts out of a salad. Nearly three-quarters of food workers believed that they knew how to "guarantee" a safe meal. Most states do not require that food providers attend educational programs, and there are no national requirements.

Sampson, acutely aware of the risks facing food-allergy sufferers, is now trying to work out a way to help desensitize people. To do this, he is relying on the idea behind the hygiene hypothesis and some of Lack's investigations: that exposure in small doses, in controlled circumstances, can build tolerance. He is trying to identify how the IgE antibody attaches to different proteins, and he uses this knowledge to have foods cooked in a way that would make the proteins less allergenic. Researchers at Mount Sinai observed, for example, that baking caused milk proteins to change shape in a way that could be less provocative to the immune system. An allergic person might be able to eat the altered proteins and become tolerant of them in all their forms. Sampson and other researchers have also configured an experimental vaccine that contains fragments of peanut protein that might "reeducate" the immune system of allergic people. Safety studies of the experimental vaccine are under way at the Jaffe Institute.

In 2008, when Maya Konoff was seven, her mother enrolled her in a research study being conducted by Dr. Sampson at the Jaffe Food Allergy Institute, funded by the NIH. She was given allergens in an altered form, and if she achieved tolerance she would be given foods that contained the allergen in its more natural state.

The treatment rooms at the institute are painted in soft tones, and the hallways are decorated with large photographs of fruits. The institute has a spotless stainless-steel kitchen; all the refrigerators and cabinets are kept locked. Diego Baraona, the chef, prepares the foods. When I visited, he showed me a batch of small muffins he had baked, with applesauce and milk, and cups of rice pudding tightly sealed in plastic. With a nurse and Jill Mindlin at Maya's side, the child was given a muffin. Maya tentatively took a bite, waited, and seemed to have no reaction. In short order, she ate the rest of the muffin. "It was very exciting for our family," Mindlin recalled, "because it meant that she was one of those kids whose bodies didn't recognize the milk protein when it was broken down in baking, so now she had potential to eat baked foods."

The next step was to try a taste of pizza. Maya took her first bite, waited, smiled, and then took another two bites. "I knew right then that things were not going well, even though Maya had not exhibited any physical symptoms," Mindlin said. "She had been so giddy, riding off the high of eating the muffin, happy and chattering, and then all of a sudden there was this pall that came over her." Maya soon broke out in hives and began vomiting. Sampson gave her an epinephrine injection. As the drug took effect, the anaphylactic reaction was arrested.

According to the protocol, Maya was supposed to come back in six months. Dr. Sampson counseled that in the meantime she should eat baked foods that included milk. When she returned, an intravenous line was inserted and an epinephrine injection pen was placed at the bedside before Maya was offered a slice of the same pizza. "It was nothing less than miraculous," her mother told me. "She ate the entire slice of pizza." Maya was observed for several hours and then given a bowl of rice pudding. The doctors told

Mindlin to expect a reaction. "But instead she ate the whole bowl of rice pudding and was fine. She jumped two levels, just by eating muffins every day," Mindlin said.

Maya returned to Mount Sinai the next day for a glass of milk. "That didn't go quite as well," Mindlin said. As Maya finished drinking, her nose began to run and she vomited. The allergic reaction was mild enough to be treated with Benadryl. When I spoke to Mindlin in December, she told me that Maya can now eat macaroni and cheese but that she is still unable to drink milk. "Even if she never progresses past this, I have no regrets about being in the study, because now she can go to a birthday party and have a slice of pizza. It's huge."

PART TWO

Animals

CARL ZIMMER

The Long, Curious, Extravagant Evolution of Feathers

FROM *National Geographic*

MOST OF US will never get to see nature's greatest marvels in person. We won't get a glimpse of a colossal squid's eye, as big as a basketball. The closest we'll get to a narwhal's unicornlike tusk is a photograph. But there is one natural wonder that just about all of us can see, simply by stepping outside: dinosaurs using their feathers to fly.

Birds are so common, even in the most paved-over places on Earth, that it's easy to take for granted both their dinosaur heritage and the ingenious plumage that keeps them aloft. To withstand the force of the oncoming air, a flight feather is shaped asymmetrically, the leading edge thin and stiff, the trailing edge long and flexible. To generate lift, a bird has merely to tilt its wings, adjusting the flow of air below and above them.

Airplane wings exploit some of the same aerodynamic tricks. But a bird wing is vastly more sophisticated than anything composed of sheet metal and rivets. From a central feather shaft extends a series of slender barbs, each sprouting smaller barbules, like branches from a bough, lined with tiny hooks. When these grasp the hooklets of neighboring barbules, they create a structural network that's feather light but remarkably strong. When a bird preens its feathers to clean them, the barbs effortlessly separate, then slip back into place.

The origin of this wonderful mechanism is one of evolution's most durable mysteries. In 1861, just two years after Darwin published *Origin of Species,* quarry workers in Germany unearthed spectacular fossils of a crow-size bird, dubbed *Archaeopteryx,* that lived about 150 million years ago. It had feathers and other traits of living birds but also vestiges of a reptilian past, such as teeth in its mouth, claws on its wings, and a long, bony tail. Like fossils of whales with legs, *Archaeopteryx* seemed to capture a moment in a critical evolutionary metamorphosis. "It is a grand case for me," Darwin confided to a friend.

The case would have been even grander if paleontologists could have found a more ancient creature endowed with more primitive feathers—something they searched for in vain for most of the next century and a half. In the meantime, other scientists sought to illuminate the origin of feathers by examining the scales of modern reptiles, the closest living relatives of birds. Both scales and feathers are flat. So perhaps the scales of the birds' ancestors had stretched out, generation after generation. Later their edges could have frayed and split, turning them into the first true feathers.

It made sense too that this change occurred as an adaptation for flight. Imagine the ancestors of birds as small, scaly, four-legged reptiles living in forest canopies, leaping from tree to tree. If their scales had grown longer, they would have provided more and more lift, allowing the protobirds to glide a little farther, then a little farther still. Only later might their arms have evolved into wings they could push up and down, transforming them from gliders to true powered fliers. In short, the evolution of feathers would have happened along with the evolution of flight.

This feathers-led-to-flight notion began to unravel in the 1970s, when the Yale University paleontologist John Ostrom noted striking similarities between the skeletons of birds and those of terrestrial dinosaurs called theropods, a group that includes marquee monsters like *Tyrannosaurus rex* and *Velociraptor.* Clearly, Ostrom argued, birds were the living descendants of theropods. Still, many known theropods had big legs, short arms, and stout, long tails—hardly the anatomy one would expect on a creature leaping from trees. Other paleontologists argued that birds did not evolve from dinosaurs—rather, their similarities derived from a shared common ancestor deeper in the past.

In 1996 Chinese paleontologists delivered startling support for Ostrom's hypothesis. It was the fossil of a small, short-armed

125-million-year-old theropod, *Sinosauropteryx*, which had one extraordinary feature: a layer of thin, hollow filaments covering its back and tail. At last there was evidence of truly primitive feathers —found on a ground-running theropod. In short, the origin of feathers may have had nothing to do with the origin of flight.

Soon paleontologists were finding hundreds of feathered theropods. With so many fossils to compare, they began piecing together a more detailed history of the feather. First came simple filaments. Later, different lineages of theropods evolved various kinds of feathers, some resembling the fluffy down on birds today, some having symmetrically arranged barbs. Other theropods sported long, stiff ribbons or broad filaments, unlike the feathers on any living birds.

The long, hollow filaments on theropods posed a puzzle. If they were early feathers, how had they evolved from flat scales? Fortunately, there are theropods with threadlike feathers alive today: baby birds. All the feathers on a developing chick begin as bristles rising up from its skin; only later do they split open into more complex shapes. In the bird embryo these bristles erupt from tiny patches of skin cells called placodes. A ring of fast-growing cells on the top of the placode builds a cylindrical wall that becomes a bristle.

Reptiles have placodes too. But in a reptile embryo each placode switches on genes that cause only the skin cells on the back edge of the placode to grow, eventually forming scales. In the late 1990s Richard Prum of Yale University and Alan Brush of the University of Connecticut developed the idea that the transition from scales to feathers might have depended on a simple switch in the wiring of the genetic commands inside placodes, causing their cells to grow vertically through the skin rather than horizontally. In other words, feathers were not merely a variation on a theme: they were using the same genetic instruments to play a whole new kind of music. Once the first filaments had evolved, only minor modifications would have been required to produce increasingly elaborate feathers.

Until recently it was thought that feathers first appeared in an early member of the lineage of theropods that leads to birds. In 2009, however, Chinese scientists announced the discovery of a bristly-backed creature, *Tianyulong*, on the ornithischian branch of the dinosaur family tree—about as distant a relative of theropods as a dinosaur can be. This raised the astonishing possibility that the ancestor of *all* dinosaurs had hairlike feathers and that some species lost them later in evolution. The origin of feathers could be pushed back further still if the "fuzz" found on some pterosaurs is confirmed to be feathers, since these flying reptiles share an even older ancestor with dinosaurs.

There's an even more astonishing possibility. The closest living relatives of birds, dinosaurs, and pterosaurs are crocodilians. Although these scaly beasts obviously do not have feathers today, the discovery of the same gene in alligators that is involved in building feathers in birds suggests that perhaps their ancestors did, 250 million years ago, before the lineages diverged. So perhaps the question to ask, say some scientists, is not how birds got their feathers, but how alligators lost theirs.

If feathers did not evolve first for flight, what other advantage could they have provided the creatures that had them? Some paleontologists have argued that feathers could have started out as insulation. Theropods have been found with their forelimbs spread over nests, and they may have been using feathers to shelter their young.

Another hypothesis has gained strength in recent years: that feathers first evolved to be seen. Feathers on birds today come in a huge range of colors and patterns, with iridescent sheens and brilliant streaks and splashes. In some cases their beauty serves to attract the opposite sex. A peacock unfolds his iridescent train, for instance, to attract a peahen. The possibility that theropods evolved feathers for some kind of display got a big boost in 2009, when scientists began to take a closer look at their structure. They discovered microscopic sacs inside the feathers, called melanosomes, that correspond precisely in shape to structures associated with specific colors in the feathers of living birds. The melanosomes are so well preserved that scientists can actually reconstruct the color of dinosaur feathers. *Sinosauropteryx*'s tail, for example, appears to have had reddish and white stripes. Perhaps the males of the species flashed their handsome tails when courting females. Or perhaps both sexes used their stripes the way zebras use theirs —to recognize their own kind or confuse predators.

*

Whatever the original purpose of feathers, they were probably around for millions of years before a single lineage of dinosaurs began to use them for flight. Paleontologists are now carefully studying the closest theropod relatives of birds for clues to how this transition occurred. One of the most revealing is a recently discovered wonder called *Anchiornis,* more than 150 million years old. The size of a chicken, it had arm feathers with black-and-white portions, creating the spangled pattern you might see on a prize rooster at a county fair. On its head it wore a gaudy rufous crown. In structure, *Anchiornis*'s plumes were nearly identical to flight feathers, except that they were symmetrical rather than asymmetrical. Without a thin, stiff leading edge, they may have been too weak for flight.

What the plumes lacked in strength, however, they made up for in number. *Anchiornis* had an embarrassment of feathers. They sprouted from its arms, legs, and even its toes. It's possible that sexual selection drove the evolution of this extravagant plumage, much as it drives the evolution of peacock trains today. And just as their long, heavy trains pose a burden to peacocks, the extravagant feathers of *Anchiornis* may have been a bit of a drag, literally.

Corwin Sullivan and his colleagues at the Institute of Vertebrate Paleontology and Paleoanthropology in Beijing have found a way that *Anchiornis* could have overcome this problem. In the theropods that were closely related to living birds, a particular wrist bone was wedge-shaped, allowing them to bend their hands. *Anchiornis*'s wrist bone was so wedge-shaped that it could fold its arms to its sides, keeping its arm feathers off the ground as it walked. Modern birds use a similar bone in flight, drawing their wings toward their bodies during an upstroke. If Sullivan and his colleagues are right, this crucial flight feature evolved long before birds took wing. It's an example of what evolutionary biologists call exaptation: borrowing an old body part for a new job. It now looks as if bird flight was made possible by a whole string of such exaptations stretching across millions of years, long before flight itself arose.

The way in which that final transition occurred continues to inspire lively debate. Some scientists argue that feathered dinosaurs evolved flight from the ground up, flapping their feathered arms as they ran. Others challenge this notion, pointing out that the "leg wings" on *Anchiornis* and other close relatives of birds would have made for very clumsy running. These researchers are reviving the old idea that protobirds used feathers to help them leap from trees, glide, and finally fly.

Ground up, trees down—why not both? Flight did not evolve in a two-dimensional world, argues Ken Dial, a flight researcher at the University of Montana, Missoula. Dial has shown that in many species a chick flaps its rudimentary wings to gain traction as it runs from predators up steep inclines, like tree trunks and cliffs. But flapping also helps steady the chick's inevitable return to lower terrain. As the young bird matures, such controlled descent gradually gives way to powered flight. Perhaps, says Dial, the path the chick takes in development retraces the one its lineage followed in evolution—winging it, so to speak, until it finally took wing.

THOMAS HAYDEN

How to Hatch a Dinosaur

FROM *Wired*

PEOPLE HAVE TOLD Jack Horner he's crazy before, but he has a knack for turning out to be right. In 1982, on the strength of seven years of undergraduate study, a stint in the Marines, and a gig as a paleontology researcher at Princeton, Horner got a job at Montana State University's Museum of the Rockies in Bozeman. He was hired as a curator but soon told his bosses that he wanted to teach paleontology. "They said it wasn't going to happen," Horner recalls. Four years and a MacArthur genius grant later, "they told me to do whatever I wanted to." Horner, sixty-five, continues to work at the museum, now filled with his discoveries. He still doesn't have a college degree.

When he was a kid in the 1950s, dinosaurs were thought to have been mostly cold, solitary, reptilian beasts—true monsters. Horner didn't agree with this picture. He saw in their hundreds-of-millions-of-years-old skeletons hints of sociability, of animals that lived in herds, unlike modern reptiles. Then in the 1970s, Horner and his friend Bob Makela excavated one of the most spectacular dinosaur finds ever—a massive communal nesting site of duck-billed dinosaurs in northwestern Montana complete with fossilized adults, juveniles, and eggs. There they found proof of crazy idea number one: the parents at the site cared for their young. Judging by their skeletons, the baby duckbills would have been too feeble to forage on their own.

Horner went on to find evidence suggesting that once hatched, the animals were fast-growing (crazy idea number two) and possibly warm-blooded (that would be three), and he continues to be at the forefront of the search for ancient bits of organic matter surviving intact in fossils (number four). Add in his work as a technical consultant on the *Jurassic Park* movies, and Horner has probably done more to shape the way we currently think about dinosaurs than any other living paleontologist.

All of which means that people are more cautious about calling him crazy these days, even when he tells them what he plans to do next: Jack Horner wants to make a dinosaur. Not from scratch—don't be ridiculous. He says he's going to do it by reverse-evolving a chicken. "It's crazy," Horner admits. "But it's also possible."

Over the past several decades, paleontologists—including Horner—have found ample evidence to prove that modern birds are the descendants of dinosaurs, everything from the way they lay eggs in nests to the details of their bone anatomy. In fact, there are so many similarities that most scientists now agree that birds actually *are* dinosaurs, most closely related to two-legged meat-eating theropods like *Tyrannosaurus rex* and velociraptor.

But "closely related" means something different to evolutionary biologists than it does to, say, the people who write incest laws. It's all relative: human beings are almost indistinguishable, genetically speaking, from chimpanzees, but at that scale we're also pretty hard to tell apart from bats.

Hints of long-extinct creatures, echoes of evolution past, occasionally emerge in real life—they're called atavisms, rare cases of individuals born with characteristic features of their evolutionary antecedents. Whales are sometimes born with appendages reminiscent of hind limbs. Human babies sometimes enter the world with fur, extra nipples, or, very rarely, a true tail. Horner's plan, in essence, is to start off by creating experimental atavisms in the lab. Activate enough ancestral characteristics in a single chicken, he reasons, and you'll end up with something close enough to the ancestor to deserve a "-saurus." At least, that's what he pitched at this year's TED Conference, the annual technology, entertainment, and design gathering held in Long Beach, California. "When I was growing up in Montana, I had two dreams," he told the crowd. "I wanted to be a paleontologist, a dinosaur paleontologist—and I wanted to have a pet dinosaur."

Already researchers have found tantalizing clues that at least some ancient dinosaur characteristics can be reactivated. Horner is the first to admit that he doesn't know enough to do the work himself, so he's actively seeking a developmental biology postdoctoral fellow to join his lab group in Montana. Horner has the big ideas, and he has some seed funding.

Now all he needed to make it happen, he told his TED audience, was a few breakthroughs in developmental biology and genetics and all the chicken eggs he could get his hands on. "What we're trying to do is take our chicken, modify it, and make," he said, "a chickenosaurus."

Horner's effort to reverse-evolve a dinosaur is not how most people envision *T. rex* making a comeback. That scientific scenario was essentially the premise of Michael Crichton's *Jurassic Park*—namely that bloodsucking insects trapped in prehistoric amber could contain enough dinosaur DNA for scientists to clone the great beasts. Horner threw himself into assessing this idea after the book came out in 1990, and he was hired as a consultant on the film trilogy. He ultimately concluded that DNA breaks down too fast in amber and in bones (no matter how exquisitely well preserved). In other words, dinosaur cloning was not feasible. But Horner hadn't given up on owning a dinosaur just yet. "I didn't really think we could do it," he says, "until I had a much better understanding of what it was that we couldn't do."

So he started reading developmental biology papers. And in 2005 he read a book called *Endless Forms Most Beautiful* by Sean Carroll. In the 1980s, Carroll helped lay the groundwork for the field of evolutionary developmental biology—evo devo—which focuses on figuring out the molecular mechanisms of evolution. It's a basic fact of biology that living things change over generations, shaped by the randomness of genetic mutation and the winnowing effects of the environment. The biologists wanted to determine what, exactly, changes. Using fruit flies, they established that just a handful of genes—most famously the homeotic, or Hox, genes—control the basic framework of a fruit fly's body. Even more surprising, those Hox genes are found in everything from nematode worms to humans, with a nearly identical sequence of amino acids called the homeodomain.

These regulatory genes—the master switches of development—contain the recipes for making certain proteins that stick to different stretches of the genome, where they function like brake shoes, controlling at what time during development, and in what part of the body, other genes (for things like growth-factor proteins or actual structural elements) get turned on. The same basic molecular components get deployed to make the six-legged architecture of an insect or fish fins or elephant trunks. Different body shapes aren't the result of different genes, though genetic makeup certainly plays a role in evolution. They're the result of different *uses* of genes during development.

So making a chicken egg hatch a baby dinosaur should really just be an issue of erasing what evolution has done to make a chicken. "There are twenty-five years of developmental biology underlying the work that makes Horner's thought experiment possible," says Carroll, a molecular biologist at the University of Wisconsin, Madison. Every cell of a turkey carries the blueprints for making a tyrannosaurus, but the way the plans get read changes over time as the species evolves.

All Horner had to do was learn how to control the control genes. He had spent decades studying fossilized dinosaur embryos, tracking in minute detail the structural and cellular changes in their skeletons as they grew. Now he immersed himself in what biologists had figured out about the molecular control of those changes. Horner reads scientific papers the way he hunts for fossils—scanning a barren landscape for rare bits of useful material—and he has found enough of them to feel optimistic.

Horner is a big man—six feet three and over two hundred pounds. It's a tight squeeze to reach the desk in his cluttered basement office at the Museum of the Rockies. Surrounded by four large LED monitors, Horner rummages among awards, family photos, and what looks like a triceratops horn in a canvas shopping bag before he finds what he's looking for: a mounted chicken skeleton. "The skeletons of a chicken and a *T. rex* really are very similar," he says. "We're going to focus on just a few of the major differences." He points out the ten or so vertebrae, several of them fused and kinked upward, that pass for the tail on a chicken. Two-legged dinosaurs had long, dramatic tails, held up from the ground to counterbalance the body. Fixing the tail will be the first step.

Step two: the hands. Many dinosaurs had two or three fingers, with sharp claws used for grasping and tearing. Birds have a "hand" at the end of each wing, but the three digits are tiny and fused together. The trick will be unfusing them. Step three will be replacing the chicken's tough keratin beak with long rows of pointy dinosaur teeth. "That is one good reason to do this in a chicken instead of an ostrich," says Horner, whose deadpan humor comes in a slow, easy-to-miss burn. "You want something small enough to catch."

He didn't actually know how to do any of this, of course. The breakthrough came in a bar. Horner doesn't remember exactly where—paleontologists tend to travel a lot—but he thinks it was in 2005. He was talking with Hans Larsson, a young Canadian paleontologist who had recently started teaching at McGill University; Horner had known him since Larsson was a graduate student at the University of Chicago. Larsson was interested in how dinosaurs lost their tails along the evolutionary road. "As soon as he started talking about looking for the genes that were responsible, I said, 'Well, if you could find those, we could just reverse the whole process.'" Larsson was thirty-four at the time and as trim and energetic as Horner is burly and unhurried—a velociraptor to the older man's triceratops. He was taken aback but didn't dismiss the idea out of hand.

Larsson is a fairly unusual paleontologist in that he studies living animals as well as fossils. He trained in paleontology and biology and today splits his time between dig sites in the Arctic (and elsewhere) and an advanced developmental biology lab. "I became a little bit dissatisfied with just pure paleontology," Larsson says. "It seemed too much like going out and collecting something, adding it to the museum drawer, and not actually testing anything." It's a frustration that every student of extinct animals has to face sooner or later: you can't keep the darned things in a lab and do experiments on them. But because of the principle of genetic conservation—the idea that all living creatures carry a substrate of very similar DNA—Larsson can study chickens, alligators, and even mice to gain insight into dinosaurs.

That work got under way in 2008—in part thanks to Horner, who donated the money to fund a postdoc in Larsson's lab for a year. The first task was to spend several years developing exquisitely sensitive techniques to follow the activity of four key regulatory networks. One of these pathways includes a gene known as Sonic Hedgehog, which controls the proliferation of cells. Another is involved with wing outgrowth. The third helps establish a top-to-bottom axis in developing limbs, and the last controls skeletal patterning. Most of these activities can be manipulated—suppressed or even stopped—using pharmacological agents. Or you can just inject more of the protein that a particular gene makes, increasing its effect. "Our plan is to start working with this toolkit and manipulate it in different parts of the embryo," Larsson says.

Like Horner, Larsson is focused on the tail and wing for now. But he wants to learn how dinosaurs became birds, not turn back the evolutionary clock. That's just Horner's crazy idea.

In 2002, Matthew Harris sat down to dissect a chicken embryo. A grad student in developmental biology at the University of Wisconsin, Harris was trying to figure out how feathers evolved. As is common practice in his field, he had turned to a deformed animal for clues; figuring out what went wrong often shows what's supposed to go right. He was working with a talpid^2, a particularly odd strain of mutant chicken best known for grotesque forelimbs and feet that can sprout as many as ten digits each—so many that a fully developed chick can't muster the biomechanical wherewithal to break out of its shell and hatch. Harris was looking beyond those obvious alterations, searching for oddities in skin, scales, and feathers.

It was one of several old specimens, collected by his PhD adviser, John Fallon, years before, right at the point of hatching. Preserved in thick, syrupy glycerol, the embryo had become nearly transparent. "I brought it out of the jar to look at it, and the outer beak, the rhamphotheca, started to come off," Harris says. "I peeled it back and then stopped—the specimen was smiling back at me." Scores of scientists had studied talpid^2 embryos for years, but Harris saw what no one else had: a neat row of pointy, uniform structures running along the jawline, hidden beneath the hard outer beak. The bird had a mouthful of toothlike buds.

Harris and his colleagues soon discovered that by stimulating production of a protein called beta-catenin in chick embryos, they could get normal nonmutants to produce neat rows of conical, crocodile-like tooth buds along their upper and lower beaks. "Chicks have the potential to create toothlike structures," Harris says. "They just need the right signal to come through."

Where Harris—now on the faculty at Harvard Medical School—saw an interesting bit of developmental biology, Horner saw yet another stepping stone to his dinosaur. The beta-catenin trick made growing chickenosaurus teeth relatively easy. Unfortunately for Horner, Harris is among those who don't see the path quite as clearly. "I respect him and what he does," Harris says. "But I think what he's trying to sell is a little outlandish."

Those chick's teeth were evidence, Harris says, that evolution had preserved the basic developmental mechanisms for starting to make teeth. But they were mere buds, with none of the design and material flourishes that make mature teeth into tearers of flesh and crushers of bone. "Development has the capacity to remake a lot of things," Harris says. "But what you lose are some of the last bits, like enamel and dentine, that are specific for teeth. You can't even find a gene for enamel in the chicken genome."

Carroll, the evo devo expert, shares that skepticism. He has done plenty of body-changing experiments on insects, manipulating the order and structure of development, and let's just say that the resulting bugs are never happy. "It's not like a Mr. Potato Head, where you just give it a tail and new hands and voilà: dinosaur," Carroll says. "That tail has got to work with the rest of the body. There's likely going to be some wiring problems, some coordination problems. Maybe some other body parts won't develop normally." He doesn't disparage the imagination behind the idea and thinks that with enough money and time Horner might get something done. But "even if you raised an adult chicken with teeth, you'd really end up with nothing more than Foghorn Leghorn with teeth," Carroll says. "And shitty teeth at that."

Horner's quest to make a dinosaur reflects what he sees as a broader problem in paleontology: digging bones out of the ground has produced huge amounts of information about prehistoric life, but he has begun to think that scientists have learned just about everything they can about dinosaurs from that method. "We'll get little chunks of DNA, and we'll figure out what colors they were," Horner says. "But the fossil record is pretty limited."

Having spent a career shaking up paleontology, Horner seems perfectly happy with the idea that even *considering* a chickenosaurus shakes up biologists. "Paleontology is ossified," says Nathan Myhrvold, the former Microsoft CTO who now dabbles in a bunch of different sciences and has worked extensively with Horner. "The methods haven't changed substantially in a hundred years."

Yes, researchers know more about dinosaurs and other extinct creatures now than they did a century ago—and Myhrvold has been a coauthor of several academic papers that contribute to that supply of knowledge. But he sees Horner's work as the first real push to bring the tools and insights of molecular and developmental biology into the paleontological fold. "Normally, paleontologists go out and walk around until they find fossils," Myhrvold says. "But it turns out that there's a place to look that's just as good as the badlands of Montana, and that's the genome of living relatives."

And if Horner is right, do we get the joy of real dinosaurs menacing the San Diego suburbs? "A lot of people say, 'You worked on *Jurassic Park,* you should know better,'" Horner says with a laugh. "But contrary to Steven Spielberg's movies, animals don't want to get even with us. We actually could have dinosaurs running around and they wouldn't be any worse than grizzly bears and mountain lions." That might seem like scant reassurance to those who spend less time wandering the badlands than Horner does. But for now, Horner has no intention of letting any of his experiments hatch. (Just give him a few years and some funding.) And because he intends only to manipulate developmental signals, without altering any DNA, any offspring of a chickenosaurus would be a normal-looking chicken. So what could possibly go wrong?

One project, if it ever happens, could give us an idea. In 2008, researchers at Penn State announced that they'd sequenced most of the genome of the woolly mammoth, extinct for 10,000 years, from samples of its hair. That prompted the Harvard geneticist George Church to claim that for around $10 million he could resurrect the mammoth. He'd take a skin cell from an elephant, even more closely related to mammoths than humans are to chimps, and then reprogram the elephantine bits of its genome into something more mammothy. Convert that into an embryo and bring it to term in an elephant uterus. No problem.

If Church were ever to try it—and there are no signs that he will—the project would have a few advantages over Horner's. DNA can last for around 100,000 years, so researchers actually have mostly intact genetic material from mammoths, avoiding the *Jurassic Park* degraded-DNA problem. And from a genetic perspective, elephants are practically mammoths already, whereas chickens have diverged pretty significantly from, say, a velociraptor. But the important point is that the technology to do this kind of work didn't exist ten years ago. It's now possible, for example, to make thousands of modifications to the genome in a single cell. Genomics has gone from an artisanal craft to something more akin to the mechanical looms of the early industrial revolution. Sure, to realize his reverse-evolution dream, Horner needs to take the technology even further. But the trend lines do seem to point in the right direction.

Back in his office, he picks up a heavy introductory developmental bio textbook from his desk. "All these books are about flies," Horner says, arching his eyebrows. "Flies are great. They're very interesting, and you can learn a lot by studying them. But . . ." He tosses the book onto a chair and stands up, walks down a long hallway to his crammed collection room and a drawer filled with every imaginable sort of bird skull—a toucan with its giant orange bill, a parrot's hooklike mouth, the flattened beak of a spoonbill. "Birds are pretty amazing, too," he says.

Developmental biologists talk about the regulatory machinery they study as a biological toolkit, a small set of mechanisms and processes that evolution uses to construct new and wonderful bodies. "Well," Horner says, "they've found the toolkit. But what good is a toolkit if you don't use it to build something?"

MICHAEL BEHAR

Faster. Higher. Squeakier.

FROM *Outside*

BACK IN THE EARLY 1960s, when the architect Louis Kahn designed the airy layout of the Salk Institute—a collection of stark concrete towers aligned like teetering dominoes on a Pacific Ocean bluff in La Jolla, California—he oriented the buildings so that robust sea breezes would waft through the upper floors. But as I descend four flights of stairs to enter a sprawling subterranean lab, the sweet ocean air turns sour. Researchers at Salk are conducting cutting-edge experiments in genetics, biology, neuroscience, and human physiology. At the core of this futuristic work are six thousand old-fashioned, defecating rodents, stacked in shoebox-size plastic cages, creating an odor far too potent for Kahn's ingenious ventilation scheme to handle.

Despite the funk, the facility is meticulously clean. Wearing powder-blue scrubs, a surgical mask, a bouffant cap, and cloth shoe covers, I enter through a sterile clean room closed off between double doors. A whitewashed hallway adjoins various smaller labs, where some mice are being injected with performance-enhancing compounds and forced to sprint on tiny treadmills. Others have had bits of their DNA reprogrammed to make them better runners. There are paunchy mice gorging on high-fat diets and svelte mice getting low-cal meals. Hunched over a metal table, a technician sorts through a squirming posse, plucking out prime studs for breeding and banishing aggressive males to solitary confinement. Mice are sacrificed and their muscles examined. Blood is sampled, hearts are inspected, kidneys and livers prodded.

This busy little world is the multimillion-dollar endeavor of Ron

Evans, a sixty-one-year-old molecular and developmental biologist who's trying to crack the code of human endurance. With help from a team of thirty-five scientists, Evans has an ambitious goal: to develop the first-ever performance-enhancing drug that can radically boost physical endurance in humans.

The "exercise in a pill" project began during the summer of 2007, when Evans made a stunning announcement. While investigating obesity, he stumbled upon a genetic switch that unexpectedly turned his lab rodents into superathletes. In August 2008, Evans published the findings in *Cell,* a prestigious scientific journal, claiming that in some cases his augmented mice could run 90 percent farther than ordinary critters. By comparison, it's considered extraordinary when a human athlete's performance jumps by only 3 percent. Evans's breakthrough would be like transforming a dawdling weekend jogger into an Ironman contender overnight. And, as Evans assures me, "This wouldn't require you to actually exercise muscle to gain a benefit."

In the now famous *Cell* paper, Evans and his coauthors—a collaborative multinational team based at research institutes in California, Massachusetts, and South Korea—confidently announced that they had found a way "to enhance training adaptation or even to increase endurance without exercise." Physiologists who'd spent their careers deconstructing the sophisticated mechanics of exercise and its numerous benefits were skeptical, dismissing the notion of pill-popping your daily workout as ludicrous.

But that didn't stop every major media outlet—including the big four networks, cable news channels, the *New York Times,* and the *Wall Street Journal*—from declaring the breakthrough a "couch potato's dream." *Nova scienceNow,* a PBS program, interviewed Evans, who said that "the benefit of exercise alone and the benefit of the drug [are] almost exact" and predicted that athletes would be the earliest adopters.

Though it may be years before doctors are writing prescriptions that turbocharge your training, serious people are aiming at that goal. Evans's group is a frontrunner in the race, but there are others: independent teams around the world developing naturally derived and synthetically engineered compounds that in preliminary animal experiments—and a few human tests—have measurably increased overall fitness.

Obviously, there will be hurdles. One is convincing biotech

firms to back the costly studies required to create a marketable drug. Another is the U.S. Food and Drug Administration, which won't green-light a new treatment that exists solely to help people run farther. (Scientists would first have to show that the drug can cure a real disease.) Even so, Evans believes that we're heading toward an inevitable day in which a pill will supplement and, in many cases, entirely replace exercise.

I first heard about Evans on the *NBC Nightly News,* shortly after slogging through a forty-minute treadmill run at my gym. When a smirking Brian Williams flashed the onscreen headline EXERCISE IN A PILL, my bullshit meter redlined. So I phoned Evans, who amiably assured me that his research was legit and invited me to visit his lab, where I could see his supermice firsthand.

Now, over the course of an introductory two-hour chat in his oak-paneled fifth-floor office, Evans, a southern California native who's tan and slim and looks far younger than his age, does his best to simplify the science. When it comes to genetics and pharmacology—subjects I've covered for more than a decade—I'm usually a quick study. Not so today. Listening to Evans delve into the complexities of cellular nutrient transfer makes my brain hurt.

Evans is goateed and wears frameless specs, designer jeans, a crisp blue oxford shirt, and black retro sneakers. On the windows, across the glass, he's scribbled elaborate equations that almost completely obscure the ocean view. Academic honors in elegant frames crowd the walls, with overflow awards aligned neatly along baseboards. On a shelf are three bobbleheads—one of Evans beside James Watson and Francis Crick, the legendary scientists who in 1962 shared a Nobel Prize with Maurice Wilkins for mapping the structure of DNA. There's a stainless-steel yo-yo on his desk and a half-empty bottle of Jose Cuervo on a coffee table. I ask about the tequila, but Evans, a wicked tennis player and avid swimmer, can't remember how it got there and would rather talk about Lance Armstrong's quads.

To be an endurance athlete like Armstrong, Evans explains, your leg muscles need lots of slow-twitch fibers. "Energy is stored in the chemical form of ATP, adenosine triphosphate," he says. "The mitochondria, the powerhouses of the cells, break down sugar and fat to create ATP." Every endurance athlete knows what comes next: when ATP stores run dry, you bonk, hit the wall—*kablooey.*

Exercise creates more slow-twitch fibers and fuels a process known as mitochondrial biogenesis. Put simply, train hard and your mitochondria multiply like microbes. More mitochondria equals more ATP and, *whoosh,* you're running sub-three-hour marathons. Among exercise physiologists, the consensus has always been that the only way to increase mitochondria was through intense, prolonged physical activity.

"Endurance is a matter of real-time generation of ATP, and it was thought that exercise was the only way to get the system to work better," says Evans, who accepted this idea until 1998. That's when he began exploring the role of genes in obesity, homing in on a genetic switch called peroxisome proliferator-activated receptor delta, or PPAR-delta, a protein known to regulate metabolism and fat burning. When your body demands fuel, PPAR-delta can influence whether it chooses glucose (sugar) or lipids (fats).

At rest, PPAR-delta is dormant. But during exercise it awakens to sustain a metabolic chain reaction that produces muscle fibers with slow-twitch properties, which feed on body fat. Vigorous exercise isn't an option if you're morbidly obese, though. So Evans wondered: what if we exercised the gene and not the muscle? Activate PPAR-delta, his thinking went, and fat-eating slow-twitch fibers would materialize like blades of grass sprouting from a freshly watered lawn.

In his first experiment, Evans coded the PPAR-delta gene to activate only in fat cells, where he thought it would have the most impact on weight loss. "We reengineered PPAR-delta in mice to be permanently on, like a light switch," he says. "What happened was a bit of a miracle. The animals slimmed down and were resistant to weight gain even on a high-fat diet." Fat cells in the mice had become more oxidative, similar to what happens when you blow air over smoldering coals and they erupt into flames. The cells could, quite literally, vaporize excess blubber.

Impressive results, but Evans wasn't satisfied. By 2004 he'd figured out how to tweak the PPAR-delta gene to fire in muscle cells. If the muscle became oxidative, as it did in the fat-cell experiment, it would cultivate the growth of mitochondria-rich slow-twitch fibers, essential for endurance.

Recalling all this, Evans grins broadly, eager to reveal the outcome. "We got marathon mice—an entire strain of animals that had become long-distance runners without ever having had to run," he says. "We proved that endurance could be genetically engineered through this particular switch. And the switch stayed on and could be passed on as a genetic trait. You could have a whole lineage of long-distance-running mice."

While we talk, Evans sits cross-legged in a sage-colored lounge chair, fiddling with pencil-thin paper wands that resemble giant chopsticks. He makes them by rolling together discarded Post-its. "Humans and spotted hyenas are endurance predators. They wear their prey out," he says, delving into a tangential discussion of fast-twitch muscle fibers in primates. I nudge him back on topic. "So we wanted to find a drug that could activate the PPAR-delta switch by injection or pill," he says, "because genetic engineering is impractical."

At this point Evans leaps from his chair and starts pacing in front of a large whiteboard. He grabs a red marker and draws a box. Inside he writes GW1516. "This is a Glaxo compound," he says, referring to the pharmaceutical giant GlaxoSmithKline, which, Evans learned, had created GW1516 more than a decade ago, later making it publicly available for biotech researchers. "They were developing it to trigger the PPAR-delta switch, because they had observed that in obese primates it tripled HDL levels, the good cholesterol." Glaxo test subjects had been receiving GW1516 in intermittent doses—enough to increase HDL but not a lot else. GW1516 was available commercially, so Evans ordered up a batch and fed it to his mice every day for five weeks, a dose that far exceeded amounts given in any previous experiments. "The effect was huge!" he says.

It sure was. Couch-potato mice could eke out a lame two-thirds of a mile. The same was true for mice given GW1516 that didn't train. Mice that didn't get GW1516 but did ten-minute daily stints on a treadmill eventually hit 1.1 miles. But mice that had both —training and GW1516—easily hit 2.3 miles.

In short, the drug had doubled the normal performance-enhancing effect of regular endurance training. Unlike mice with genetically altered PPAR-delta, GW1516 had no impact on sedentary animals. Exercise, it seemed, was an essential part of the equation, though Evans didn't know why.

He submitted the results to *Cell* in 2007. But the editors wanted more and initially refused to publish his paper. "We had ended the story with a drug working in the context of exercise, and the *Cell* reviewers said, 'Look, you can't leave us hanging, because if what you're saying is correct, then the real breakthrough would be to completely replace exercise.' They wanted us to take it to the next level, to find a drug that could enhance performance without *any* exercise. That was something nobody had done before, and we didn't think it was possible."

Evans persisted, searching for another substance to flip the PPAR-delta switch. The winner was a chemical compound called AICAR (pronounced *aye-car*), which had been around since the 1980s and was being used in clinical trials for the treatment of ischemic reperfusion, a rare complication of coronary bypass surgery that occurs when blood flow restored to previously damaged arteries causes inflammation and damage to heart tissue.

"We knew AICAR could stimulate a more oxidative metabolism," Evan says. "There were reports that it had been given to people, and activity in muscle had been measured. But these studies were all based on single injections. They weren't giving it once a day for thirty days. When we did that, the results were beautiful."

Once again, here was an experimental compound readily available to scientists—but one that nobody had thought to test in a high-dose way. Mice that hadn't done any exercise but were given AICAR could run 23 percent longer and 44 percent farther than sedentary mice that didn't get the drug.

Sure, it wasn't the doubling of endurance seen with GW1516. But the AICAR mice hadn't trained at all. They'd become remarkably fit by doing nothing.

Once word got out about AICAR and GW1516, Evans figured that human athletes would jump the gun and start ingesting the stuff. Before Evans published his *Cell* paper, he tipped off the World Anti-Doping Agency (WADA), the Montreal-headquartered outfit that sets drug-testing and enforcement policies adopted by every Olympic and many non-Olympic sports. WADA asked him to devise a test to detect the drugs in urine and blood and added both compounds to its list of banned substances. It didn't take long for the drug to make news: the French Anti-Doping Agency alleged that AICAR had been used by riders in the 2009 Tour de France, though it never came forward with specific allegations or named names.

Meanwhile, on supplement-oriented Web forums like RxMuscle.com, the buzz grew quickly. "I can't wait!" one poster declared. "Give me some of that GW1516!" Another wrote: "AICAR is already available on the grey market." There's also an online clearinghouse, aicar.co.uk, which provides AICAR data and calls the compound "a new dawn in dieting and fitness . . . the revolutionary AICAR and GW1516 are the newest buddies of athletes."

Other studies have shown that a healthy abundance of mitochondria can mitigate aging and make it easier to lose weight, factors that will likely extend AICAR and GW1516 use well beyond a handful of zealous endurance athletes. And as Evans points out, "These compounds are easy to make or obtain." He shows me a Web site where a licensed research institute can buy GW1516 online; AICAR is also available from biotech suppliers. "Type 'purchase AICAR' into a search engine," Evans suggests. I quickly find some, though it's not cheap: a thousand bucks for ten grams, about twenty times the street price of cocaine.

Though AICAR is easy to buy, that doesn't mean it's safe. "The big problem with AICAR is the side effects," says Laurie Goodyear, an associate professor of medicine at Harvard Medical School and a senior investigator at the Joslin Diabetes Center. "Athletes would get a huge increase in lactic acid. There's also a molecular mutation in the heart that can lead to sudden death. Certainly there's a possibility that drugs could be developed to increase endurance. But I don't believe AICAR would improve performance in humans." In 2008 Goodyear wrote an article for the *New England Journal of Medicine* that examined Evans's claims. Her parting advice: "Don't get too comfortable on that couch just yet."

In addition, Evans's mice were couch potatoes that had never exercised. With a fitness baseline of zero, there's plenty of room to improve. "If you have a highly trained athlete that already has high levels of mitochondria," Goodyear says, "it's possible they may get some benefit, but I don't think it would be really huge."

Mark Davis, who directs the Exercise Biochemistry Laboratory at the University of South Carolina, believes that in elite athletes mitochondria hit a ceiling at some point, in part because "too many of them can actually be toxic to the cells."

Evans isn't dissuaded, but he's also aware that the FDA won't approve any drug unless it has a specific disease application. So his team is focusing its resources and funding on identifying legitimate therapeutic uses for AICAR and GW1516. He's been talking with biotech firms about funding clinical trials that "target frailty, or people in wheelchairs who can't exercise, or who've gone through surgery and are bedridden." There's also potential for treating diabetes, high cholesterol, obesity, metabolic disorders, and muscular dystrophy. Still, Evans isn't bashful about admitting where the real money will be made.

"If approved, this can be prescribed by doctors for anything you want," he says. "And very few people in this country get the recommended minimum of forty minutes a day of exercise. So when you ask me who would want a drug that confers some of the benefits of exercise without actually exercising, it would be the majority of the population."

That's the kind of market pharmaceutical companies love—and it's why Evans isn't the only one dreaming of riches. "We're doing the same thing with resveratrol as Evans did with AICAR," says Johan Auwerx, a professor of energy metabolism at École Polytechnique Fédérale de Lausanne, in Switzerland. "There is a healthy competition going on between us." Resveratrol, in case you missed its being touted on *Oprah, 60 Minutes,* and *Good Morning America,* is a potent antioxidant found in the skin of red grapes. In mice given colossal doses—to match them, you'd have to chug something like 50,000 bottles of wine a day—it curbed aging, lowered blood sugar, slowed the spread of cancer, and spawned mitochondria.

"Our mice ran longer when we gave them resveratrol," says Auwerx, who is now trying to identify other natural compounds more potent still, to be taken as over-the-counter supplements.

Evans is one of only a few scientists targeting endurance through genes—and that, he believes, gives him an edge. "A lot of people study the end result [of exercise] or study hormones," he says. "But what controls everything is the genome. It's the heart of the entire system, and it's what I'm interested in changing." First, though, he must test hordes of mice for every conceivable side effect, inject them with varying dosages of AICAR at different intervals to establish an optimal treatment program, and demonstrate that his compounds can do something other than just endow rodents (and ultimately humans) with superlative endurance. There's also that minor little discrepancy between rodent and human physiology: after all, the list of prototype miracle drugs that performed spectacularly in mice and then failed catastrophically during human clinical trials is long and sordid.

In the basement lab at the Salk Institute, my escort—a bespectacled postdoc with a boyish smile, named Vihang Narkar—raises another concern. Athletes rely on mental stamina as much as on physical fortitude to push through pain, a phenomenon that could make it tricky to accurately assess the potency of AICAR or GW1516 in people. Our tendency to either persevere or succumb is inextricably tied to both brain and brawn. But according to Narkar, mice bonk for only one reason: their muscles are simply depleted of every last bit of ATP.

While chatting with Narkar, I sort of forget about the "wild" mice we've left on the treadmill, which he'd set up earlier to demonstrate a typical training session. Neither has been tainted with the magic jock-juice, and they appear identical—just two ordinary plump and furry rodents with no discerning features that might hint at physical prowess. They've been plodding along for twenty minutes or so without much fuss.

Then Narkar ups the belt speed to 18 meters per minute (roughly two-thirds of a mile per hour) and the mice burst into a gallop. He pushes it higher, to 22 meters per minute, and the mouse closest to me takes the lead—a born athlete, for sure—while the slower mouse languishes. Suddenly, the speedy mouse dashes right off the end of the belt, springs from the treadmill, plummets four feet to the floor, and is headed in a blind sprint for the door when Narkar nabs it with a lightning-fast lunge-and-swipe combo that he's definitely performed more than once.

I insist that Narkar mistakenly grabbed an AICAR mouse for this demonstration. "Some wild mice are just inherently better runners," he says. It's apropos that he recognizes its natural athleticism—often the game-changing wild card integral to competitive sports—since he's part of a team developing drugs that could give any beer-bellied schlub a fast-track ticket to the peloton.

BIJAL P. TRIVEDI

The Wipeout Gene

FROM *Scientific American*

OUTSIDE TAPACHULA, CHIAPAS, Mexico—ten miles from Guatemala. To reach the cages, we follow the main highway out of town, driving past soy, cocoa, banana, and lustrous dark green mango plantations thriving in the rich volcanic soil. Past the tiny village of Rio Florido the road degenerates into an undulating dirt track. We bump along on waves of baked mud until we reach a security checkpoint, guard at the ready. A sign posted on the barbed wire–enclosed compound pictures a mosquito flanked by a man and a woman: *Estos mosquitos genéticamente modificados requieren un manejo especial,* it reads. *We play by the rules.*

Inside, cashew trees frame a cluster of gauzy mesh cages perched on a platform. The cages hold thousands of *Aedes aegypti* mosquitoes—the local species, smaller and quieter than the typical buzzing specimens found in the States. At seven A.M., the scene looks ethereal: rays of sunlight filter through layers of mesh, creating a glowing, yellow hue. Inside the cages, however, genetically modified mosquitoes are waging a death match against the locals, an attempted genocide-by-mating that has the potential to wipe out dengue fever, one of the world's most troublesome, aggressive diseases.

Throughout a swath of subtropical and tropical countries, four closely related dengue viruses infect about 100 million people annually, causing a spectrum of illness—from flulike aches to internal hemorrhaging, shock, and death. No vaccine or cure exists. As with other mosquito-borne diseases, the primary public health strategy is to prevent people from being bitten. To that end, authorities attempt to rid neighborhoods of standing water where the insects breed, spray with insecticides, and distribute bed nets and other low-tech mosquito blockers. They pursue containment, not conquest.

Anthony James, however, is mounting an offensive. James, a molecular biologist at the University of California, Irvine, and his colleagues have added genes to *A. aegypti* that block the development of flight muscles in females. When a genetically modified male mosquito mates with a wild female, he passes his engineered genes to the offspring. The females—the biters—don't survive long. When they emerge from the pupal stage, they sit motionless on the water. They won't fly, mate, or spread disease. The male progeny, in contrast, will live to spread their filicidal seed. In time, the absence of female offspring should lead to a population crash, which James's collaborator has already demonstrated in the controlled environment of an indoor laboratory in Colorado. Now he has brought his bugs south.

The technology marks the first time scientists have genetically engineered an organism to specifically wipe out a native population to block disease transmission. If the modified mosquitoes triumph, then releasing them in dengue-endemic zones worldwide could prevent tens of millions of people from suffering. Yet opponents of the plan warn of unintended consequences—even if mosquitoes are the intended victims.

Researchers also struggle with how to test their creations. No international laws or agencies exist to police trials of new transgenic organisms. For the most part, scientists and biotech companies can do what they want—even performing uncontrolled releases of test organisms in developing countries, neither warning the residents that their backyards are about to become a de facto biocolonialist field laboratory nor gaining their consent.

James has spent years attempting to play it straight. He has worked with community leaders in Tapachula, acquiring property through the traditional land-sharing program and building a secure test facility—all arduous, time-consuming, careful work. But he is not the only researcher testing modified mosquitoes outside the lab. James's colleague Luke Alphey, founder of the UK-based biotechnology company Oxitec, has quietly pursued a more aggressive test strategy. In 2009 and 2010 his organization took advantage of the minimal regulations in the Caribbean's Grand Cay-

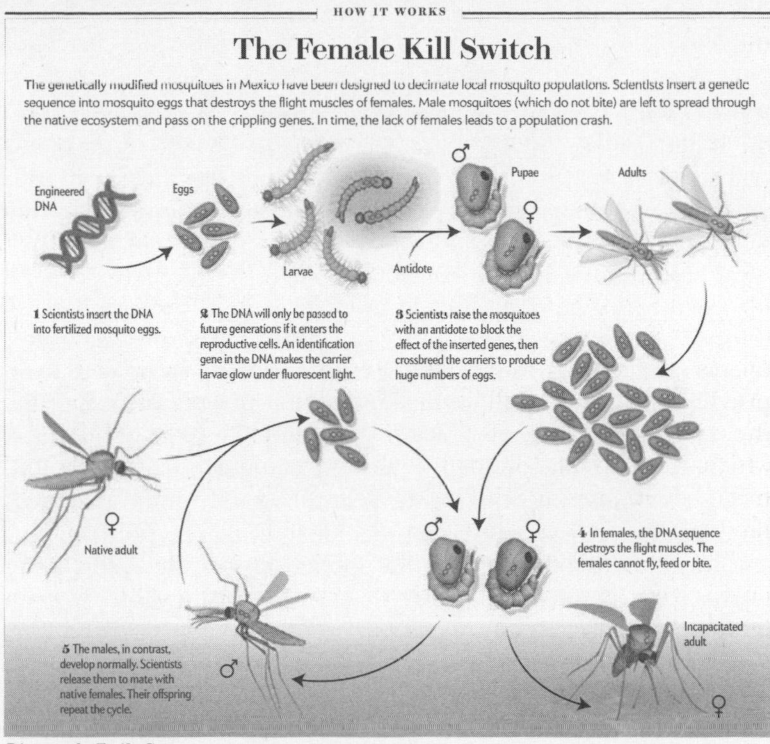

HOW IT WORKS
The Female Kill Switch
The genetically modified mosquitoes in Mexico have been designed to decimate local mosquito populations. Scientists insert a genetic sequence into mosquito eggs that destroys the flight muscles of females. Male mosquitoes (which do not bite) are left to spread through the native ecosystem and pass on the crippling genes. In time, the lack of females leads to a population crash.
Engineered DNA
Eggs
Larvae
Antidote
Pupae
♂
♀
Adults
1 Scientists insert the DNA into fertilized mosquito eggs.
2 The DNA will only be passed to future generations if it enters the reproductive cells. An identification gene in the DNA makes the carrier larvae glow under fluorescent light.
3 Scientists raise the mosquitoes with an antidote to block the effect of the inserted genes, then crossbreed the carriers to produce huge numbers of eggs.
♀
Native adult
♂
♀
4 In females, the DNA sequence destroys the flight muscles. The females cannot fly, feed or bite.
Incapacitated adult
5 The males, in contrast, develop normally. Scientists release them to mate with native females. Their offspring repeat the cycle.
♂
♀

Diagram by Emily Cooper

man Island to release millions of genetically modified mosquitoes into the wild. James first learned of the experiments when Alphey described them publicly at a conference in Atlanta in 2010—fourteen months after the fact. Since then, Oxitec has continued the trials, releasing modified mosquitoes in Malaysia and Brazil.

Experts fear Oxitec's actions could trigger a backlash against all genetically modified insects, reminiscent of Europe's rejection of GM crops, a move that could snuff out the technology before scientists can fully understand both its promise and its potential consequences.

That would be a shame, because the technology has such promise. The Colorado lab test demonstrated that the modified mosquitoes work in a controlled environment, although a few indoor cages are not the wilds of Central America, Brazil, or Malaysia. To fight the sickness and death that ride inside the mosquito, the scientists' creations must conquer the jungle.

Forced Sterilization

In 2001 James was already a pioneer of modern molecular mosquito genetics: the first researcher to genetically alter a mosquito and the first to clone a mosquito gene. That year he decided to apply his knowledge to the problems of disease transmission. He wondered if he could use a strategy designed to control agricultural pests on mosquitoes instead.

A year before, Alphey, then at the University of Oxford, had developed a technique for generating fruit flies harboring genes that selectively killed females. The population-control strategy is just a postgenomic riff on sterile insect technology (SIT), which has successfully controlled crop pests for sixty years. Technicians rear vast numbers of insects, sterilizing the males with blasts of radiation. When they mate with females in local fields, the union produces no offspring. The strategy is insecticide-free, targets only the pest species, and has been successfully applied many times—including a large-scale Mediterranean fruit fly (Medfly) eradication program in 1977 in Tapachula.

Unfortunately, sterile insect technology has never worked with mosquitoes. Radiation severely weakens adult males, and the processes of sorting and transport kill them before they can mate. Extending Alphey's new fruit fly technique to mosquitoes, however, would enable researchers to design effectively sterile male mosquitoes from the genome up.

To kill female mosquitoes—the ones that suck blood and spread disease—James needed to hijack a genetic region that only females make use of. In 2002 James and Alphey identified a naturally occurring switch that controls flight-muscle development in females. Turn it off, and flight muscles won't develop. Female mosquitoes emerging from the pupal stage just squat on the water's surface, flightless, unable to attract mates. It was the perfect target.

Alphey founded Oxitec in 2002 to capitalize on the technology. In 2005 the Foundation for the National Institutes of Health, funded in large part by the Bill & Melinda Gates Foundation, granted James $20 million to test genetic strategies against dengue. James gave Oxitec $5 million to build the mosquitoes.

The collaborators designed a stretch of DNA that included a handful of genes and the regulatory switches needed to turn them on and off at the correct time. The system works like a relay team. During the mosquito's metamorphosis from larva to adult, the female-specific switch flips on, activating the first gene, which produces a protein. This protein activates a second switch that kicks on gene number two, which then manufactures a toxin that destroys the female's flight muscles. The researchers also added genes for fluorescent proteins that make modified larvae glow red and green, allowing them to monitor the spread of the genes through the population.

To breed large populations of a mosquito that they had explicitly programmed to die, Alphey and James needed a way to protect the females from the toxic gene cassette until after they reproduced. The trick was lacing the water with an antidote—the antibiotic tetracycline, which blocks production of the flight muscle–destroying protein. This design is also an emergency fail-safe: if a few of these genetically modified mosquitoes escape, they cannot reproduce without the drug.

The first tests of the new breed came in 2008 and 2009, when Megan Wise de Valdez, a colleague of James's who at the time was based at Colorado State University, introduced modified males to a population of ordinary *A. aegypti* mosquitoes in the laboratory. Within five months the population crashed. The kill switch worked. The next step was to bring the modified mosquitoes into the field.

Breakbone Fever

In Tapachula, where James has set up his netted laboratory, dengue has long been a problem, as it has been in much of Mexico. "Dengue is my most important concern on a day-to-day basis," said Hermilo Domínguez Zárate, undersecretary of health for Chiapas, when I visited the region last year. Dengue spreads explosively, causing the most hardship in densely populated areas.

During my trip to Chiapas I toured Pobres Unidos—Poor United—an impoverished neighborhood on Tapachula's outskirts that suffered the most dengue cases in 2009 and 2010, along with Janine Ramsey, a parasitologist on James's team who leads day-to-day work at the field site, and Rogelio Danis-Lozano, a medical epidemiologist.

One home we visited belonged to Maria, who asked that I not

use her last name. As with most homes in Pobres Unidos, Maria's house has only three walls, like a house on a movie set, so she has no way to keep mosquitoes out. The moist dirt floor creates a humid environment that lures the insects close. Piles of trash and dozens of containers collect rainwater, providing countless locations for mosquitoes to deposit eggs.

Danis-Lozano directed our attention to a large yellow tub brimming with fresh water and pointed to hundreds of skinny, black, threadlike mosquito larvae swimming vigorously in erratic zigzag patterns. Maria knows about dengue, of course, but Danis-Lozano discovered she had no idea that the larvae in her washtub morph into disease-spreading mosquitoes.

It is a scene that is mirrored in poor, crowded neighborhoods worldwide. More than one hundred countries suffer from dengue, from Asia to Africa to the Americas. Symptoms of dengue's mild form—"breakbone fever"—mimic the flu: fever, joint and muscle pain, and crippling headaches that last about a week. A second infection can trigger potentially deadly dengue hemorrhagic fever, which induces vomiting, severe abdominal cramps, and internal hemorrhaging. Blood streams from the eyes, nose, mouth, and vagina. Without treatment, hemorrhagic dengue kills up to 20 percent of its victims; with costly expert care, mortality drops to 1 percent. The annual worldwide death toll exceeds that of all other viral hemorrhagic fevers—including Ebola and Marburg—combined.

In 2008 the epidemiologist David M. Morens and Anthony S. Fauci, director of the National Institute of Allergy and Infectious Diseases, warned that dengue is "one of the world's most aggressive reemerging infections." The frequency and magnitude of outbreaks have been rising, spread by growing international travel and the exodus of people to cities. Caseloads have doubled every decade since the 1970s. In 2009 Florida public health officials reported the first dengue cases there in more than seven decades, raising fears among epidemiologists that the disease would soon take root in the continental United States.

One reason James decided to apply his genetic technology to the fight against dengue fever—instead of, say, malaria—is that the virus is primarily transmitted by a single species of mosquito. (Between thirty and forty species of mosquito carry malaria.) *A. aegypti,* the world's main dengue vector, is an invasive, tree-dwelling African species that hitched a ride on slave ships some four hundred years ago. It is now an urbanite, breeding beside homes in anything that holds a few tablespoons of clean water. The mosquito bites during the day, so bed nets provide no protection. And it bites humans almost exclusively, drawing the nutrients that give it a life span of up to a month—plenty of time to bite and spread disease.

A. aegypti is stealthy, lacking the sharp, unnerving buzz that provokes a swift swat or panicked wave. Inside the secure insectary at the Regional Center for Public Health Research in Tapachula, I could barely hear a swarm of transgenic mosquitoes in a small cage. Laura Valerio, an entomologist at the University of California, Davis, stuck her gloved hand inside to point out a female. The intrusion scared the males, which took flight and zoomed around the cage. Females, however, just sat there or hopped away clumsily.

Modified mosquito larvae would later be moved to James's field site, which consists of five pairs of cages, each with a control cage housing a population of wild mosquitoes and a treatment cage where modified mosquitoes mix with locals. Each cage is guarded by multiple layers of mesh—protection against escapees—which researchers must carefully navigate through as they add new test subjects to the experiment.

The strict protocol is an attempt to avoid past errors. Developing countries have long made a convenient location for First World field trials, but a cavalier attitude toward the local environment has led to backlash that derailed entire research programs. Perhaps no field is more fraught with abuses—both real and perceived—than genetically modified organisms.

Poison in the Wells

In 1969, for example, the World Health Organization and the Indian government teamed up to study genetic control of three mosquito species: *Culex fatigans,* which spreads filariae (parasites that cause elephantitis); *A. aegypti,* which spreads dengue and yellow fever; and malaria-spreading *Anopheles stephensi.* The US government funded some of the research.

In 1972 a scientist anonymously published an article in India's *National Herald* alleging that researchers had been placing mosquitoes treated with thiotepa—described as a mustard gas derivative that causes birth defects and cancer in animals—in village drinking wells. The scientists in charge of the project issued a timid rebuttal and rebuffed subsequent interview requests from the press. Then in 1974 the Press Trust of India ran a story with the incendiary headline "WHO Works for U.S. Secret Research in India." The article alleged that the mosquito project was being used to test the practicality of using *A. aegypti* as a biowarfare agent. India was being used to test "chemicals or methods not permitted in sponsoring countries," the account ran, also charging that *A. aegypti* was being studied because "its eggs (unlike those of other mosquitoes) can be dried, put on a piece of paper in an envelope and mailed to any part of the country where they can hatch." Although the investigators strenuously denied the allegations, the public relations debacle prompted the WHO to abandon the program.

Since then, investigators have been terrified of conducting field trials of genetically modified (GM) organisms, says Stephanie James (no relation to Anthony), director of the Grand Challenges in Global Health initiative at the Foundation for the National Institutes of Health. "There was a real psychological barrier. They knew they couldn't afford to mess up."

"All my career I've been told you'll never get people to agree to do this," Anthony James told me. At the 2005 inaugural dinner for Grand Challenges grant recipients, he consulted Jim Lavery, who specializes in the science of community engagement at Toronto's Center for Global Health Research at St. Michael's Hospital. "GM freaks people out," James said. "So how do you involve the community?"

Lavery suggested choosing a location where dengue was a significant public health issue and control methods were failing, in a country with a stringent, sophisticated regulatory structure capable of assessing the risks and benefits of a genetically modified, dengue-fighting mosquito. That way locals would be comfortable that the effort would not endanger or exploit them. He and the mosquito field-trial veteran Thomas Scott of U.C. Davis helped Anthony James assemble an international team of mosquito ecologists, anthropologists, and ethicists long before he had enough mosquitoes to test.

By 2006 Tapachula was the frontrunner for these trials. Mexico had national laws on genetically modified organisms and had signed the Cartagena Protocol on Biosafety—the international framework for importing them. Experience with the medfly meant the Tapachula community wasn't "freaked out" by the idea of modifying an insect, Lavery says.

"At first the request for land sounded strange," said Martimiano Barrios Matute, leader of the farming community where the experiment is based. Why would anyone want to build large cages and fill them with man-made mosquitoes? The community was also confused about what transgenic mosquitoes could do. Could escapees hurt them or their fields? Would their sterility be transferred to other insects?

James and his group addressed the community's concerns and purchased the land to build the cages through the traditional communal land-ownership program in the area. And they continue to engage the locals as the experiment continues.

In a weekly town hall gathering in the Casa de la Cultura on Tapachula's historic main square, Ramsey, the project's field site manager, described the project to an audience of community leaders, thirty men and five women. It was hard to tell that she was an American expatriate as she held the room transfixed; she was animated, gesturing, and joking.

When she concluded, the audience cautiously asked questions. One man asked if he could visit the mosquito cages. Another wanted to know what happens if mosquitoes escape. A young woman asked why people are against transgenics. An elderly man from a mountain village asked whether malaria and dengue are different. Ramsey answered them all, then smiled and shook hands as she left.

"Now that we understand, even more so do we like the project," said Barrios Matute, a slender soy farmer with gold-capped teeth. "It will benefit not only Rio Florido but all around Rio Florido and Mexico and other parts of the world."

The Great Escape

While all this slow scientific and community work was going on in Mexico, Alphey was quietly taking a dramatically different approach. Last November he arrived at the annual meeting of the American Society of Tropical Medicine and Hygiene with a surprising story to tell. Beginning in September 2009, Alphey said, Oxitec had been releasing genetically modified mosquitoes on

Grand Cayman Island in the Caribbean. (The mosquitoes are similar to the ones being tested in Tapachula, but not identical —in the Cayman strain, both male and female mosquitoes die as larvae.) Between May and October of 2010, Oxitec released more than 3 million male mosquitoes, he revealed, which cut the indigenous *A. aegypti* population by 80 percent. The data have been submitted for publication.

Alphey defended his gung-ho approach, saying that Oxitec leaves outreach largely to the governments because they understand the cultural sensitivities. On Grand Cayman, outreach involved one five-minute spot on the local nightly news broadcast and a pamphlet that described the mosquitoes as sterile, avoiding any mention of genetic modification. There were no public meetings or opportunities for residents to voice concerns.

Alphey justified his actions at the Atlanta meeting. "In terms of publicity, we were only doing it in the Cayman Islands," he said. "We only need the community, people on the island, to know about it."

Mark Q. Benedict, a molecular biologist at the University of Perugia in Italy and a consultant to the Gates Foundation, says Oxitec has broken no laws and calls the Cayman trials "courageous" for testing technology bound to attract "attention, both good and bad." Benedict says confused and conflicting media reports created the impression of "the lone scientist who rushes out with his bucket of mosquitoes and throws them into the environment without any oversight. That is not happening." Oxitec works with both local and national governments to gain approval before any field test.

Still, the Cayman release has provoked strong emotions—distrust, disappointment, and frustration—from many of Alphey's colleagues, environmental groups, and the public. "The international community was taken by surprise that this release had happened," says Bart Knols, a medical entomologist at the University of Amsterdam and managing director of MalariaWorld. "Now the outside world perceives Oxitec as secretive, which makes the public wonder why. It breeds suspicion."

This is promising technology, Knols says. "If some party messes up badly and misinforms the public, the risk is that other GM trials will suffer." Now, because of Oxitec, he adds, "we have the same problems as the WHO had in India in 1976."

Other experts say the company is preying on countries with minimal bureaucracy and regulations. In the Cayman Islands, Oxitec conducted its trials in a place with a "streamlined regulatory structure," says Stephanie James, where the ink was barely dry on a biosafety bill that has yet to become law.

Malaysia was next. Amid protests from twenty-plus nonprofit organizations, Oxitec launched a trial in an uninhabited area last December. A follow-up in a nearby village is pending. Even with a newly minted National Biosafety Board that monitors modified organisms and the 2009 Malaysian Biosafety Act regulations, many feel that Malaysia lacks the experience to monitor the experiment, says Gurmit Singh, chair of the nonprofit Center for Environment, Technology and Development, Malaysia.

Anthony James slumped in a chair as we discussed the situation but, always diplomatic, said flatly, "That's the difficulty of working with corporations. I can't control corporate partners." He added, "If it blows up, I told you so. If not, you got lucky." James said that Oxitec's approach would be impossible in Mexico, adding that he is confident his team's community engagement activities have "set a standard for testing genetically modified organisms."

Alphey is undeterred. Earlier this year Oxitec launched a six-month trial in a poor suburb of Juazeiro, Bahia, in northern Brazil, which is plagued by mosquitoes and dengue year-round. Later this year Alphey plans to return to Grand Cayman to pit the Tapachula and Cayman strains of transgenic mosquitoes against the local mosquitoes to see which lives longer, flies farther, and is better at mating with local females. Mosquito-control officials in Panama and the Philippines have shown interest, as have the authorities in Florida.

Permanent Spread

Of course, many groups oppose the release of any transgenic organisms, no matter how thoughtfully the scientists explain themselves beforehand. Janet Cotter, a senior scientist at Greenpeace Research Laboratories, warns that "Oxitec's release of GM mosquitoes is extremely risky. There's no such thing as one hundred percent sterility, so there are going to be some fertile females that will be released, and we don't know the implications of that."

Some people wonder if it is ethical—or safe—to eliminate an

organism, even in just a small geographic area. Proponents argue that *A. aegypti* is an invasive species that has evolved to exploit a solely human niche. "Urban *A. aegypti* is not part of any significant food chain," says Phil Lounibos, a mosquito ecologist at the Florida Medical Entomology Laboratory. Yet Lounibos doubts whether eliminating *A. aegypti* would stop dengue transmission permanently. "A previous campaign to eradicate this species from the Americas in the 1950s and 1960s, when it was the primary vector of urban yellow fever, failed miserably," he says. The invasive Asian tiger mosquito—another good dengue vector—readily occupies niches vacated by *A. aegypti.* Moreover, both the Cayman and Tapachula mosquito strains, even if successful, are not permanent. Migration of mosquitoes from neighboring regions into Tapachula could foil eradication attempts and mandate frequent releases of the modified males to keep the population in check.

James and his collaborators have been developing a self-sustaining but more controversial solution. It uses a "gene drive system," which promotes the spread of dengue-resistance genes through a wild mosquito population, blocking the replication of at least one form of the dengue virus, known as type 2. Unlike the Tapachula mosquitoes, which die soon after release, mosquitoes outfitted with a gene drive will persist in the environment. James says field trials for gene drive systems are still a few years away.

"Something that spreads genes through populations is going to have much more difficult regulatory hurdles," James says, "so I'm happy to take something that is self-limiting, not sustainable, like [the Tapachula strain] and have that be our first shot."

Undersecretary of Health Domínguez Zárate views the genetically modified approach as "low cost and high creativity." "If dengue was something with less importance, then why modify something from nature?" he asks. "We need to respect nature as much as we can." Still, the costs of dengue outweigh the potential environmental risks, he says. "It's worth the gamble."

SY MONTGOMERY

Deep Intellect

FROM *Orion*

ON AN UNSEASONABLY warm day in the middle of March, I traveled from New Hampshire to the moist, dim sanctuary of the New England Aquarium, hoping to touch an alternate reality. I came to meet Athena, the aquarium's forty-pound, five-foot-long, two-and-a-half-year-old giant Pacific octopus.

For me, it was a momentous occasion. I have always loved octopuses. No sci-fi alien is so startlingly strange. Here is someone who, even if she grows to one hundred pounds and stretches more than eight feet long, could still squeeze her boneless body through an opening the size of an orange; an animal whose eight arms are covered with thousands of suckers that taste as well as feel; a mollusk with a beak like a parrot and venom like a snake and a tongue covered with teeth; a creature who can shape-shift, change color, and squirt ink. But most intriguing of all, recent research indicates that octopuses are remarkably intelligent.

Many times I have stood mesmerized by an aquarium tank, wondering, as I stared into the horizontal pupils of an octopus's large, prominent eyes, if she was staring back at me—and if so, what was she thinking?

Not long ago, a question like this would have seemed foolish, if not crazy. How can an octopus know anything, much less form an opinion? Octopuses are, after all, "only" invertebrates—they don't even belong with the insects, some of whom, like dragonflies and dung beetles, at least seem to show some smarts. Octopuses are classified within the invertebrates in the mollusk family, and many mollusks, like clams, have no brain.

Only recently have scientists accorded chimpanzees, so closely related to humans that we can share blood transfusions, the dignity of having a mind. But now, increasingly, researchers who study octopuses are convinced that these boneless, alien animals—creatures whose ancestors diverged from the lineage that would lead to ours roughly 500 to 700 million years ago—have developed intelligence, emotions, and individual personalities. Their findings are challenging our understanding of consciousness itself.

I had always longed to meet an octopus. Now was my chance: senior aquarist Scott Dowd arranged an introduction. In a back room, he would open the top of Athena's tank. If she consented, I could touch her. The heavy lid covering her tank separated our two worlds. One world was mine and yours, the reality of air and land, where we lumber through life governed by a backbone and constrained by jointed limbs and gravity. The other world was hers, the reality of a nearly gelatinous being breathing water and moving weightlessly through it. We think of our world as the "real" one, but Athena's is realer still: after all, most of the world is ocean, and most animals live there. Regardless of whether they live on land or water, more than 95 percent of all animals are invertebrates, like Athena.

The moment the lid was off, we reached for each other. She had already oozed from the far corner of her lair, where she had been hiding, to the top of the tank to investigate her visitor. Her eight arms boiled up, twisting, slippery, to meet mine. I plunged both my arms elbow deep into the 57-degree water. Athena's melon-sized head bobbed to the surface. Her left eye (octopuses have one dominant eye, just as humans have a dominant hand) swiveled in its socket to meet mine. "She's looking at you," Dowd said.

As we gazed into each other's eyes, Athena encircled my arms with hers, latching on with first dozens, then hundreds, of her sensitive, dexterous suckers. Each arm has more than two hundred of them. The famous naturalist and explorer William Beebe found the touch of the octopus repulsive. "I have always a struggle before I can make my hands do their duty and seize a tentacle," he confessed. But to me, Athena's suckers felt like an alien's kiss—at once a probe and a caress. Although an octopus can taste with all of its skin, in the suckers both taste and touch are exquisitely developed. Athena was tasting me and feeling me at once, knowing my skin, and possibly the blood and bone beneath, in a way I could never fathom.

When I stroked her soft head with my fingertips, she changed color beneath my touch, her ruby-flecked skin going white and smooth. This, I learned, is a sign of a relaxed octopus. An agitated giant Pacific octopus turns red, its skin gets pimply, and it erects two papillae over the eyes, which some divers say look like horns. One name for the species is devil fish. With sharp, parrot-like beaks, octopuses can bite, and most have neurotoxic, flesh-dissolving venom. The pressure from an octopus's suckers can tear flesh (one scientist calculated that to break the hold of the suckers of the much smaller common octopus would require a quarter ton of force). One volunteer who interacted with an octopus left the aquarium with arms covered in red hickeys.

Occasionally an octopus takes a dislike to someone. One of Athena's predecessors at the aquarium, Truman, felt this way about a female volunteer. Using his funnel, the siphon near the side of the head used to jet through the sea, Truman would shoot a soaking stream of salt water at this young woman whenever he got a chance. Later she quit her volunteer position for college. But when she returned to visit several months later, Truman, who hadn't squirted anyone in the meantime, took one look at her and instantly soaked her again.

Athena was remarkably gentle with me—even as she began to transfer her grip from her smaller outer suckers to the larger ones. She seemed to be slowly but steadily pulling me into her tank. Had it been big enough to accommodate my body, I would have gone in willingly. But at this point I asked Dowd if perhaps I should try to detach from some of the suckers. With his help, Athena and I pulled gently apart.

I was honored that she appeared comfortable with me. But what did she know about me that informed her opinion? When Athena looked into my eyes, what was she thinking?

While Alexa Warburton was researching her senior thesis at Middlebury College's newly created octopus lab, "every day," she said, "was a disaster."

She was working with two species: the California two-spot, with a head the size of a clementine, and the smaller Florida species *Octopus joubini.* Her objective was to study the octopuses' behavior in a T-shaped maze. But her study subjects were constantly thwarting her.

The first problem was keeping the octopuses alive. The 400-gallon tank was divided into separate compartments for each animal. But even though students hammered in dividers, the octopuses found ways to dig beneath them—and eat each other. Or they'd mate, which is equally lethal. Octopuses die after mating and laying eggs, but first they go senile, acting like people with dementia. "They swim loop-the-loop in the tank, they look all googly-eyed, they won't look you in the eye or attack prey," Warburton said. One senile octopus crawled out of the tank, squeezed into a crack in the wall, dried up, and died.

It seemed to Warburton that some of the octopuses were purposely uncooperative. To run the T-maze, the preveterinary student had to scoop an animal from its tank with a net and transfer it to a bucket. With bucket firmly covered, octopus and researcher would take the elevator down to the room with the maze. Some octopuses did not like being removed from their tanks. They would hide. They would squeeze into a corner where they couldn't be pried out. They would hold on to some object with their arms and not let go.

Some would let themselves be captured, only to use the net as a trampoline. They'd leap off the mesh and onto the floor—and then run for it. Yes, *run.* "You'd chase them under the tank, back and forth, like you were chasing a cat," Warburton said. "It's so *weird!*"

Octopuses in captivity actually escape their watery enclosures with alarming frequency. While on the move, they have been discovered on carpets, along bookshelves, in a teapot, and inside the aquarium tanks of other fish—upon whom they have usually been dining.

Even though the Middlebury octopuses were disaster prone, Warburton liked certain individuals very much. Some, she said, "would lift their arms out of the water the way dogs jump up to greet you." Though in their research papers the students refer to each octopus by a number, the students named them all. One of the *joubini* was such a problem they named her the Bitch. "Catching her for the maze always took twenty minutes," Warburton said. "She'd grip onto something and not let go. Once she got stuck in a filter and we couldn't get her out. It was awful!"

Then there was Wendy. Warburton used Wendy as part of her thesis presentation, a formal event that was videotaped. First Wendy squirted salt water at her, drenching her nice suit. Then, as Warburton tried to show how octopuses use the T-maze, Wendy scurried to the bottom of the tank and hid in the sand. Warburton says the whole debacle occurred because the octopus realized in advance what was going to happen. "Wendy," she said, "just didn't feel like being caught in the net."

Data from Warburton's experiments showed that the California two-spots quickly learned which side of a T-maze offered a terra-cotta pot to hide in. But Warburton learned far more than her experiments revealed. "Science," she says, "can only say so much. I know they watched me. I know they sometimes followed me. But they are so different from anything we normally study. How do you prove the intelligence of someone so different?"

Measuring the minds of other creatures is a perplexing problem. One yardstick scientists use is brain size, since humans have big brains. But size doesn't always match smarts. As is well known in electronics, anything can be miniaturized. Small brain size was the evidence once used to argue that birds were stupid—before some birds were proven intelligent enough to compose music, invent dance steps, ask questions, and do math.

Octopuses have the largest brains of any invertebrate. Athena's is the size of a walnut—as big as the brain of the famous African gray parrot Alex, who learned to use more than one hundred spoken words meaningfully. That's proportionally bigger than the brains of most of the largest dinosaurs.

Another measure of intelligence: you can count neurons. The common octopus has about 130 million of them in its brain. A human has 100 billion. But this is where things get weird. Three-fifths of an octopus's neurons are not in the brain; they're in its arms.

"It is as if each arm has a mind of its own," says Peter Godfrey-Smith, a diver, professor of philosophy at the Graduate Center of the City University of New York, and an admirer of octopuses. For example, researchers who cut off an octopus's arm (which the octopus can regrow) discovered that not only does the arm crawl away on its own, but if the arm meets a food item, it seizes it—and tries to pass it to where the mouth would be if the arm were still connected to its body.

"Meeting an octopus," writes Godfrey-Smith, "is like meeting an intelligent alien." Their intelligence sometimes even involves changing colors and shapes. One video online shows a mimic octopus alternately morphing into a flatfish, several sea snakes, and a lionfish by changing color, altering the texture of its skin, and shifting the position of its body. Another video shows an octopus materializing from a clump of algae. Its skin exactly matches the algae from which it seems to bloom—until it swims away.

For its color palette, the octopus uses three layers of three different types of cells near the skin's surface. The deepest layer passively reflects background light. The topmost may contain the colors yellow, red, brown, and black. The middle layer shows an array of glittering blues, greens, and golds. But how does an octopus decide what animal to mimic, what colors to turn? Scientists have no idea, especially given that octopuses are likely *colorblind.*

But new evidence suggests a breathtaking possibility. Woods Hole Marine Biological Laboratory and University of Washington researchers found that the skin of the cuttlefish *Sepia officinalis,* a color-changing cousin of octopuses, contains gene sequences usually expressed only in the light-sensing retina of the eye. In other words, cephalopods—octopuses, cuttlefish, and squid—may be able to see with their skin.

The American philosopher Thomas Nagel once wrote a famous paper titled "What Is It Like to Be a Bat?" Bats can see with sound. Like dolphins, they can locate their prey using echoes. Nagel concluded that it was impossible to know what it's like to be a bat. And a bat is a fellow mammal like us—not someone who tastes with its suckers, sees with its skin, and whose severed arms can wander about, each with a mind of its own. Nevertheless, there are researchers still working diligently to understand what it's like to be an octopus.

Jennifer Mather spent most of her time in Bermuda floating facedown on the surface of the water at the edge of the sea. Breathing through a snorkel, she was watching *Octopus vulgaris*—the common octopus. Although indeed common (they are found in tropical and temperate waters worldwide), at the time of her study in the mid-1980s, "nobody knew what they were doing."

In a relay with other students from six-thirty in the morning till six-thirty at night, Mather worked to find out. Sometimes she'd see an octopus hunting. A hunting expedition could take five minutes or three hours. The octopus would capture something, inject it with venom, and carry it home to eat. "Home," Mather found, is where octopuses spend most of their time. A home, or den, which an octopus may occupy for only a few days before switching to a new one, is a place where the shell-less octopus can safely hide: a hole in a rock, a discarded shell, or a cubbyhole in a sunken ship. One species, the Pacific red octopus, particularly likes to den in stubby brown glass beer bottles.

One octopus Mather was watching had just returned home and was cleaning the front of the den with its arms when, suddenly, it left the den, crawled a meter away, picked up one particular rock, and placed the rock in front of the den. Two minutes later, the octopus ventured forth to select a second rock. Then it chose a third. Attaching suckers to each rock, the octopus carried the load home, slid through the den opening, and carefully arranged the three objects in front. Then it went to sleep. What the octopus was thinking seemed obvious: "Three rocks are enough. Good night!"

The scene has stayed with Mather. The octopus "must have had some concept," she said, "of what it wanted to make itself feel safe enough to go to sleep." And the octopus knew how to get what it wanted: by employing foresight, planning—and perhaps even tool use. Mather is the lead author of *Octopus: The Ocean's Intelligent Invertebrate,* which includes observations of octopuses who dismantle Lego sets and open screw-top jars. Coauthor Roland Anderson reports that octopuses even learned to open the childproof caps on Extra Strength Tylenol pill bottles—a feat that eludes many humans with university degrees.

In another experiment, Anderson gave octopuses plastic pill bottles painted different shades and with different textures to see which ones evoked more interest. Usually each octopus would grasp a bottle to see if it was edible and then cast it off. But to his astonishment, Anderson saw one of the octopuses doing something striking: she was blowing carefully modulated jets of water from her funnel to send the bottle to the other end of her aquarium, where the water flow sent it back to her. She repeated the action twenty times. By the eighteenth time, Anderson was already on the phone with Mather with the news: "She's bouncing the ball!"

This octopus wasn't the only one to use the bottle as a toy. Another octopus in the study also shot water at the bottle, sending it back and forth across the water's surface, rather than circling the tank. Anderson's observations were reported in the *Journal of Comparative Psychology*. "This fit all the criteria for play behavior," said Anderson. "Only intelligent animals play—animals like crows and chimps, dogs and humans."

Aquarists who care for octopuses feel that not only can these animals play with toys, but they may *need* to play with toys. An *Octopus Enrichment Handbook* has been developed by Cincinnati's Newport Aquarium, with ideas of how to keep these creatures entertained. One suggestion is to hide food inside Mr. Potato Head and let your octopus dismantle it. At the Seattle Aquarium, giant Pacific octopuses play with a baseball-sized plastic ball that can be screwed together by twisting the two halves. Sometimes the mollusks screw the halves back together after eating the prey inside.

At the New England Aquarium, it took an engineer who worked on the design of cubic zirconium to devise a puzzle worthy of a brain like Athena's. Wilson Menashi, who began volunteering at the aquarium weekly after retiring from the Arthur D. Little Corporation sixteen years ago, devised a series of three Plexiglas cubes, each with a different latch. The smallest cube has a sliding latch that twists to lock down, like the bolt on a horse stall. Aquarist Bill Murphy puts a crab inside the clear cube and leaves the lid open. Later he lets the octopus lift the lid. Finally he locks the lid, and invariably the octopus figures out how to open it.

Next he locks the first cube within a second one. The new latch slides counterclockwise to catch on a bracket. The third box is the largest, with two different locks: a bolt that slides into position to lock down and a second one like a lever arm, sealing the lid much like the top of an old-fashioned glass canning jar.

All the octopuses Murphy has known learned fast. They typically master a box within two or three once-a-week tries. "Once they 'get it,'" he says, "they can open it very fast"—within three or four minutes. But each may use a different strategy.

George, a calm octopus, opened the boxes methodically. The impetuous Gwenevere squeezed the second-largest box so hard she broke it, leaving a hole two inches wide. Truman, Murphy said, was "an opportunist." One day, Murphy put two crabs inside the smaller of the two boxes, and the crabs started to fight. Truman was too excited to bother with locks. He poured his seven-foot-long body through the two-inch hole Gwenevere had made, and visitors looked into his exhibit to find the giant octopus squeezed, suckers flattened, into the tiny space between the walls of the fourteen-cubic-inch box outside and the six-cubic-inch one inside it. Truman stayed inside for half an hour. He never opened the inner box—probably he was too cramped.

Three weeks after I first met Athena, I returned to the aquarium to meet the man who had designed the cubes. Menashi, a quiet grandfather with a dark mustache, volunteers every Tuesday. "He has a real way with octopuses," Dowd and Murphy told me. I was eager to see how Athena behaved with him.

Murphy opened the lid of her tank, and Athena rose to the surface eagerly. A bucket with a handful of fish sat nearby. Did she rise so eagerly sensing the food? Or was it the sight of her friend that attracted her? "She knows me," Menashi answered softly.

Anderson's experiments with giant Pacific octopuses in Seattle prove Menashi was right. The study exposed eight octopuses to two unfamiliar humans, dressed identically in blue aquarium shirts. One person consistently fed a particular octopus, and another always touched it with a bristly stick. Within a week, at first sight of the people, most octopuses moved toward the feeders and away from the irritators, at whom they occasionally aimed their water-shooting funnels.

Upon seeing Menashi, Athena reached up gently and grasped his hands and arms. She flipped upside down, and he placed a capelin in some of the suckers near her mouth, at the center of her arms. The fish vanished. After she had eaten, Athena floated in the tank upside down, like a puppy asking for a belly rub. Her arms twisted lazily. I took one in my hand to feel the suckers—did that arm know it had hold of a different person than the other arms did? Her grip felt calm, relaxed. With me earlier, she had seemed playful, exploratory, excited. The way she held Menashi with her suckers seemed to me like the way a long-married couple holds hands at the movies.

I leaned over the tank to look again into her eyes, and she bobbed up to return my gaze. "She has eyelids like a person does," Menashi said. He gently slid his hand near one of her eyes, causing her to slowly wink.

Biologists have long noted the similarities between the eyes of an octopus and the eyes of a human. The Canadian zoologist N. J. Berrill called it "the single most startling feature of the whole animal kingdom" that these organs are nearly identical: both animals' eyes have transparent corneas, regulate light with iris diaphragms, and focus lenses with a ring of muscle.

Scientists are currently debating whether we and octopuses evolved eyes separately or whether a common ancestor had the makings of the eye. But intelligence is another matter. "The same thing that got them their smarts isn't the same thing that got us our smarts," says Mather, "because our two ancestors didn't *have* any smarts." Half a billion years ago, the brainiest thing on the planet had only a few neurons. Octopus and human intelligence evolved independently.

"Octopuses," writes philosopher Godfrey-Smith, "are a separate experiment in the evolution of the mind." And that, he feels, is what makes the study of the octopus mind so philosophically interesting.

The octopus mind and the human mind probably evolved for different reasons. Humans—like other vertebrates whose intelligence we recognize (parrots, elephants, and whales)—are long-lived, social beings. Most scientists agree that an important event that drove the flowering of our intelligence was our ancestors' beginning to live in social groups. Decoding and developing the many subtle relationships among our fellows, and keeping track of these changing relationships over the course of the many decades of a typical human life span, was surely a major force shaping our minds.

But octopuses are neither long-lived nor social. Athena, to my sorrow, may live only a few more months: the natural life span of a giant Pacific octopus is only three years. If the aquarium added another octopus to her tank, one might eat the other. Except to mate, most octopuses have little to do with others of their kind.

So why is the octopus so intelligent? What is its mind *for?* Mather thinks she has the answer. She believes the event driving the octopus toward intelligence was the loss of the ancestral shell. Losing the shell freed the octopus for mobility. Now they didn't need to wait for food to find them; they could hunt like tigers. And while most octopuses love crab best, they hunt and eat dozens of other species—each of which demands a different hunting strategy. Each animal you hunt may demand a different skill set. Will you camouflage yourself for a stalk-and-ambush attack? Shoot through the sea for a fast chase? Or crawl out of the water to capture escaping prey?

Losing the protective shell was a tradeoff. Just about anything big enough to eat an octopus will do so. Each species of predator also demands a different evasion strategy—from flashing warning coloration if your attacker is vulnerable to venom, to changing color and shape for camouflage, to fortifying the door to your home with rocks.

Such intelligence is not always evident in the laboratory. "In the lab, you give the animals this situation, and they react," points out Mather. But in the wild, "the octopus is actively discovering his environment, not waiting for it to hit him. The animal makes the decision to go out and get information, figures out how to get the information, gathers it, uses it, stores it. This has a great deal to do with consciousness."

So what does it feel like to be an octopus? Philosopher Godfrey-Smith has given this a great deal of thought, especially when he meets octopuses and their relatives, giant cuttlefish, on dives in his native Australia. "They come forward and look at you. They reach out to touch you with their arms," he said. "It's remarkable how little is known about them . . . but I could see it turning out that we have to change the way we think of the nature of the mind itself to take into account minds with less of a centralized self."

"I think consciousness comes in different flavors," agrees Mather. "Some may have consciousness in a way we may not be able to imagine."

*

In May I visited Athena a third time. I wanted to see if she recognized me. But how would I be able to tell? Scott Dowd opened the top of her tank for me. Athena had been in a back corner but floated immediately to the top, arms outstretched, upside down.

This time I offered her only one arm. I had injured a knee and, feeling wobbly, used my right hand to steady myself while I stood on the stool to lean over the tank. Athena in turn gripped me with only one of her arms and very few of her suckers. Her hold on me was remarkably gentle.

I was struck by this, since Murphy and others had first described Athena's personality to me as "feisty." "They earn their names," Murphy had told me. Athena is named for the Greek goddess of wisdom, war, and strategy. She is not usually a laid-back octopus, as George had been. "Athena could pull you into the tank," Murphy had warned. "She's curious about what you are."

Was she less curious now? Did she remember me? I was disappointed that she did not bob her head up to look at me. But perhaps she didn't need to. She may have known from the taste of my skin who I was. But why was this feisty octopus hanging in front of me in the water, upside down?

Then I thought I might know what she wanted from me. She was begging. Dowd asked around and learned that Athena hadn't eaten in a couple of days, then allowed me the thrilling privilege of handing her a capelin.

Perhaps I had understood something basic about what it felt like to be Athena at that moment: she was hungry. I handed a fish to one of her larger suckers, and she began to move it toward her mouth. But soon she brought more arms to the task, and covered the fish with many suckers—as if she were licking her fingers, savoring the meal.

A week after I last visited Athena, I was shocked to receive this e-mail from Scott Dowd: "Sorry to write with some sad news. Athena appears to be in her final days, or even hours. She will live on, though, through your conveyance." Later that same day, Dowd wrote to tell me that she had died. To my surprise, I found myself in tears.

Why such sorrow? I had understood from the start that octopuses don't live very long. I also knew that while Athena did seem to recognize me, I was not by any means her special friend. But she was very significant to me, both as an individual and as a representative from her octopodan world. She had given me a great gift: a deeper understanding of what it means to think, to feel, and to know. I was eager to meet more of her kind.

And so, it was with some excitement that I read this e-mail from Dowd a few weeks later: "There is a young pup octopus headed to Boston from the Pacific Northwest. Come shake hands (x8) when you can."

MARK W. MOFFETT

Ants & the Art of War

FROM *Scientific American*

THE RAGING COMBATANTS form a blur on all sides. The scale of the violence is almost incomprehensible, the battle stretching beyond my field of view. Tens of thousands sweep ahead with a suicidal single-mindedness. Utterly devoted to duty, the fighters never retreat from a confrontation—even in the face of certain death. The engagements are brief and brutal. Suddenly, three foot soldiers grab an enemy and hold it in place until one of the bigger warriors advances and cleaves the captive's body, leaving it smashed and oozing.

I back off with my camera, gasping in the humid air of the Malaysian rain forest, and remind myself that the rivals are ants, not humans. I have spent months documenting such deaths through a field camera that I use as a microscope, yet I still find it easy to forget that I am watching tiny insects—in this case, a species known as *Pheidologeton diversus,* the marauder ant.

Scientists have long known that certain kinds of ants (and termites) form tight-knit societies, with members numbering in the millions, and that these insects engage in complex behaviors. Such practices include traffic management, public health efforts, crop domestication and, perhaps most intriguingly, warfare: the concentrated engagement of group against group in which both sides risk wholesale destruction. Indeed, in contending with these issues and others, we modern humans in many respects have come to resemble ants more than our closest living relatives, the apes, which live in societies of less than two hundred. Only recently, however, have researchers begun to appreciate just how closely the war strategies of ants mirror our own. It turns out that for ants, as for humans, warfare involves an astonishing array of tactical choices about methods of attack and strategic decisions about when and where to wage war.

Shock and Awe

Remarkably, these similarities in warfare exist despite sharp differences between ants and humans in both biology and societal structure. Ant colonies consist mostly of sterile females that function as workers or soldiers, occasionally a few short-lived males that serve as drones, and one or more fertile queens. Members operate without a power hierarchy or permanent leader. Although queens are the center of colony life because they reproduce, they do not lead troops or organize labor. Rather colonies are decentralized, with workers that individually know little making combat decisions that nonetheless prove effective at the group level without oversight —a process called swarm intelligence. But although ants and humans have divergent lifestyles, they fight their foes for many of the same economic reasons, including access to dwelling spaces, territory, food, and even labor—certain ant species kidnap competitors to serve as slaves.

The tactics ants use in war depend on what is at stake. Some ants succeed in battle by being constantly on the offensive, calling to mind the Chinese military general Sun Tzu's assertion in his sixth-century B.C. book *The Art of War* that "rapidity is the essence of war." Among army ants, species of which inhabit warm regions around the world, and a few other groups, such as Asia's marauder ant, hundreds or even millions of individuals proceed blindly in a tight phalanx, attacking prey and enemies as they come across them. In Ghana I witnessed a seething carpet of workers of the army ant species *Dorylus nigricans* searching together across an area 100 feet wide. These African army ants—which, in species such as *D. nigricans* that move in broad swaths, are called driver ants—slice flesh with bladelike jaws and can make short work of victims thousands of times their size. Although vertebrate creatures can usually outrun ants, in Gabon I once saw an antelope, caught in a snare, eaten alive by a colony of driver ants. Both army ants and marauder ants will drive rival ants from food—the sheer number of troops is sufficient to overrun any rivals and control their food supply thereafter. But army ants almost always hunt en masse with a more malicious aim, storming other ant societies to seize the colony's larvae and pupae as food.

The advancing phalanxes of army and marauder ants are reminiscent of the fighting formations that humans have used from ancient Sumerian times to the regimented fronts of the American Civil War. Marching together in this way, without a specific target, as humans sometimes did, makes every raid a gamble: the ants might proceed over barren ground and find nothing. Other ant species send a far smaller number of workers called scouts out from the nest to search separately for food. By fanning out across a larger area while the rest of the colony stays home, they encounter more prey and enemies.

Yet colonies that rely on scouts may kill fewer adversaries in total because a scout must return to its nest to assemble a fighting force—usually by depositing a chemical called a pheromone for the reserve troops to follow. In the time it takes a scout to assemble those troops for battle, the enemy may have regrouped or retreated. In contrast, the workers of the army ants or marauder ants can immediately summon any help they require because a slew of assistants are marching directly behind them. The result is maximal shock and awe.

Allocating the Troops

It is not just the huge numbers of fighters that make the army and marauder ants so deadly. My research on marauder ants has shown that troops are deployed in ways that increase efficiency and reduce the cost to a colony. How an individual is deployed depends on the female's size. Marauder ant workers vary more in size than workers of any other ant species. The tiny "minor" workers (the foot soldiers of my opening description) move quickly to the front lines—the danger zone where competing ant colonies or prey are first encountered. A single minor has no more chance against the enemy than would an equally small scout of a lone-hunting species. But their sheer numbers at the front of a raid present a commanding barricade. Although some may die along the way, the minors slow or incapacitate the enemy until the larger workers, known as the medias and the majors, arrive to deliver the deathblow. The medias and the majors are much scarcer than the minors but far more lethal, with some individuals weighing five hundred times as much as one minor.

The minors' sacrifices on the front lines assure a low mortality for the medias and the majors, which require far more resources for the colony to raise and maintain. Putting the easily replaced fighters at greatest risk is a time-honored battle technique. Ancient river-valley societies did the same thing with conscripted farmers, cheaply obtained and available in droves, who absorbed the worst of the warfare. Meanwhile the elite soldiers, who received the best training and the finest weapons and armor, remained relatively safe within these hordes. And just as human armies may defeat their enemies by attrition, destroying unit by unit rather than attacking a whole force at once—a tactic known to military strategists as "defeat in detail"—so, too, do marauder ants mow down enemies a few at a time as a raid advances instead of engaging the enemy's entire strength.

In addition to killing other enemy species and prey, marauder ants intensely defend the areas around their nests and food from other colonies of their own kind. The medias and majors hang back while each minor grabs an opponent's limb. These confrontations last for hours and are deadlier than the jostles that occur between the marauder and its other competitors. Hundreds of little ants become interlocked over a few square feet as they slowly tear one another asunder.

This insect variant of hand-to-hand combat represents the common mode of killing among ants. Mortality is nearly certain, reflecting the cheapness of labor in a large colony. Ants that are less cavalier about loss of troops employ long-range weapons that allow them to hurt or impede the enemy from afar, for example, stunning their enemy with a Mace-like spray, as *Formica* wood ants from Europe and North America do, or dropping small stones onto enemy heads, as *Dorymyrmex bicolor* ants from Arizona do.

Research conducted by Nigel Franks, now at the University of Bristol in England, and his colleagues has demonstrated that the organized violence practiced by army ants and marauders is consistent with Lanchester's square law, one of the equations developed in World War I by the engineer Frederick Lanchester to understand potential strategies and tactics of opposing forces. His math showed that when many fights occur simultaneously within an arena, greater numbers trump individual fighting power. Only when the danger becomes extreme do the larger marauder ants put themselves at risk; for example, workers of all sizes will rush an entomologist foolish enough to dig up their nest, with the majors inflicting the most savage bites.

Still, just as Lanchester's square law does not apply in all situations for warring humans, neither does it describe all the behaviors of warring ants. Slave-making ants offer a fascinating exception. Certain slave makers steal the brood of their target colony to raise as slaves in the slave makers' nest. The slave makers' tough armor, or exoskeleton, as it is termed, and daggerlike jaws give them superior fighting abilities. Yet they are greatly outnumbered by the ants in the colonies they raid for slaves. To avoid being massacred, some slave makers release a "propaganda" chemical that throws the raided colony into disarray and keeps its workers from ganging up on them. In so doing, as shown by Franks and his then graduate student Lucas Partridge at the University of Bath, they are following another Lanchester strategy that at times applies also for humans. This so-called linear law holds that when battles are waged as one-on-one engagements—which is what the propaganda substance allows—victory is assured for the superior fighters even when they are outnumbered. In fact, a colony besieged by slave makers will often allow the invaders to do this plundering without any fighting or killing.

Among ants, a fighter's value to its colony bears on the risks the ant takes: the more expendable it is, the more likely it is to end up in harm's way. The guards lining marauder foraging trails, for instance, are usually elderly or maimed workers that often struggle to stay upright while lunging at intruders. As Deby Cassill of the University of South Florida reported in *Naturwissenschaften* in 2008, only older (months-old) fire ants engage in fights, whereas weeks-old workers run off, and days-old individuals feign death by lying motionless when under attack. Viewed from the ant perspective, the human practice of conscripting healthy youngsters might seem senseless. But anthropologists have found some evidence that, at least in a few cultures, successful human warriors tend to have more offspring. A reproductive edge might make combat worth the personal risk for people in their prime—an advantage unattainable by ant workers, which do not reproduce.

Territorial Control

Other humanlike military strategies emerge from observations of weaver ants. Weaver ants occupy much of the canopy of tropical forests in Africa, Asia, and Australia, where colonies may span several trees and contain 500,000 individuals—comparable to the enormous populations of some army ants. Weavers also resemble army ants in being highly aggressive. Yet the two have entirely different modi operandi. Whereas army ants do not defend territories because they stay packed together while roaming in search of other ant species to attack for food, weaver colonies are entrenched at one site, spreading their workers wide within it to keep competitors out of every inch of their turf.

They handily control huge spaces within the trees by defending a few choke points such as the spot at which the tree trunk meets the ground. Leafy "barrack nests" placed strategically in the crowns distribute the troops where they are most needed.

Weaver ant workers are also more independent than army ant workers. Army ant raids function by stripping away the workers' autonomy. Because the army ant troops confine themselves to the close quarters of their advancing pack, they require relatively few communication signals. They respond to enemies and prey in a highly regimented way. Weavers, in contrast, wander more freely and are more versatile in their response to opportunities and threats. The differences in style call to mind the contrasts between the rigidity of Frederick the Great's armies and the flexibility and mobility of Napoleon Bonaparte's troops.

Like army ants, weaver ants take similar tacks in dealing with prey and destroying an enemy: in both cases, a weaver deploys a short-range recruitment pheromone from its sternal gland to summon nearby reinforcements to make the kill. Other weaver ant communiqués are specific to warfare. When a worker returns from a fight with another colony, it jerks its body at passing ants to alert them to the ongoing combat. At the same time, it deposits a different scent along its path, a pheromone released from the rectal gland that its colony mates follow to the battlefield. Moreover, to claim a previously unoccupied space, workers will use yet another signal, defecating on the spot, much as canines mark their territory by urinating on it.

A Matter of Size

For both ants and humans, the propensity to engage in true warfare is related at least in a rough way to the size of the society. Small colonies seldom conduct protracted battles except in defense. Like human hunter-gatherers, who are often nomadic and tend to live hand to mouth, the tiniest ant societies, which contain just a few dozen individuals, do not build a fixed infrastructure of trails, food stashes, or dwelling places worth dying for. At times of intense conflict between groups, these ants, like their human counterparts, will often choose flight over fight.

Modest-sized societies will likely have more resources to defend but are still small enough to be judicious about jeopardizing their troops. Honeypot ants of the southwestern United States, which live in medium-size colonies containing a few thousand individuals, provide an example of danger mitigation. To harvest nearby prey unchallenged, a honeypot colony may stage a preemptive tournament near a neighboring nest to keep the enemy busy rather than risking deadly battles outright. During the tournament the rivals stand high on their six legs and circle one another. This "stilting" behavior mirrors the mostly bloodless, ceremonial displays of strength commonplace in small human clans, as the biologists Bert Hölldobler of Arizona State University and E. O. Wilson of Harvard University first suggested. With luck, the colony with the smaller stilting ants—typically from the weaker colony—can retreat without loss of life, but the winning side will wreak havoc on their enemies given the opportunity, devouring the losers' brood and abducting workers, called repletes, that are swollen with food that they regurgitate on request for hungry nest mates. The honeypot victors will drag the repletes back to their nest and keep these living larders as slaves. To avoid this fate, reconnaissance workers survey the tournament to assess whether their side is outnumbered and, if necessary, set in motion a retreat.

Full-bore conflicts appear to be most common for ant species with mature colonies composed of hundreds of thousands of individuals or more. Scientists have tended to consider these large social insect societies inefficient because they produce fewer new queens and males per capita than smaller groups do. I see them instead as being so productive that they have the option to invest not only in reproduction but in a workforce that exceeds the usual labor requirements—much the way our bodies invest in fatty tissue we can draw on in hard times. Different researchers have posited that individual ants have less work to do as colonies grow larger and that this leaves more of them inactive at any one time. Colony growth would thereby amplify the expansion of a dedicated army reserve that can take full advantage of Lanchester's square law in its encounters with enemies. Similarly, most anthropologists see warfare—the concentrated engagements of group against group described earlier—as having emerged only after human societies underwent a population explosion fueled by the invention of agriculture.

Superorganisms and Supercolonies

Ultimately the capacity for extreme forms of warfare in ants arises from a social unity that parallels the unity of cells in an organism. Cells recognize one another by means of chemical cues on their surface; a healthy immune system attacks any cell with different cues. In most healthy colonies, ants, too, recognize one another by means of chemical cues on their body surface, and they attack or avoid foreigners with a different scent. Ants wear this scent like a national flag tattooed on their bodies. The permanence of the scent means ant warfare can never end with one colony usurping another. Midstream switches in allegiance are impossible for adult ants. With perhaps a few rare exceptions, each worker is a part of its natal society until it dies. (Not that the interests of ant and colony always coincide. Workers of some species can attempt to reproduce and be thwarted—much as conflicts of interest between genes can occur within an organism.) This identification with their colony is all ants have, because they form anonymous societies: beyond distinguishing castes such as soldiers from queens, ant workers do not recognize one another as individuals. Their absolute social commitment is the fundamental feature of living as part of a superorganism, in which the death of a worker is of no more consequence than cutting a finger. The bigger the colony, the less a small cut is felt.

The most breathtaking example of colony allegiance in the ant world is that of the *Linepithema humile* ant. Though native to Argentina, it has spread to many other parts of the world by hitching rides on human cargo. In California the biggest of these "supercolonies" ranges from San Francisco to the Mexican border and may contain a trillion individuals, united throughout by the same "national" identity. Each month millions of Argentine ants die along battlefronts that extend for miles around San Diego, where clashes occur with three other colonies in wars that may have been going on since the species arrived in the state a century ago. The Lanchester square law applies with a vengeance in these battles. Cheap, tiny, and constantly being replaced by an inexhaustible supply of reinforcements as they fall, Argentine workers reach densities of a few million in the average suburban yard. By vastly outnumbering whatever native species they encounter, the supercolonies control absolute territories, killing every competitor they contact.

What gives these Argentines their relentless fighting ability? Many ant species, as well as some other creatures, including humans, exhibit a "dear enemy effect," in which, after a period of conflict, death rates sharply decline as the two sides settle on a boundary—often with an unoccupied no man's land between them. In the floodplains where Argentine ants originated, however, warring colonies must stop fighting each time the waters rise, forcing them to higher ground. The conflict is never settled; the battle never ends. Thus their wars continue unabated, decade after decade.

The violent expansions of ant supercolonies bring to mind how human colonial superpowers once eradicated smaller groups, from Native Americans to Australian Aborigines. Luckily, humans do not form superorganisms in the sense I have described: our allegiances can shift over time to let immigrants in, to permit nations to fluidly define themselves. Although warfare may be inescapable among many ants, it is, for us, avoidable.

PART THREE

Humans (the Good)

DEBORAH BLUM

The Scent of Your Thoughts

FROM *Scientific American*

THE MOMENT THAT STARTS Martha McClintock's scientific career is a whim of youth. Even, she recalls, a ridiculous moment. It is summer 1968, and she is a Wellesley College student attending a workshop at the Jackson Laboratory in Maine. A lunch-table gathering of established researchers is talking about how mice appear to synchronize their ovary cycles. And twenty-year-old McClintock, sitting nearby, pipes up with something like, "Well, don't you know? Women do that, too."

"I don't remember the exact words," she says now, sitting relaxed and half-amused in her well-equipped laboratory at the University of Chicago. "But everyone turned and stared." It is easy to imagine her in that distant encounter—the same direct gaze, the same friendly face and flyaway hair. Still, the lunch-table group is not charmed; it informs her that she does not know what she is talking about.

Undaunted, McClintock raises the question with some graduate students who are also attending the workshop. They bet that she will not be able to find data to support her assertion. She returns to Wellesley and talks this matter over with her undergraduate adviser, Patricia Sampson. And Sampson throws it back at her: take the bet, do the research, prove yourself right or wrong.

Three years later, now a graduate student, McClintock publishes a two-page paper entitled "Menstrual Synchrony and Suppression" in the journal *Nature.* (*Scientific American* is part of Nature Publishing Group.) It details a rather fascinating effect seen in some 135 residents of Wellesley dormitories during an academic year. In that span, menstrual cycles apparently began to shift, especially among women who spent a lot of time together. Menstruation became more synchronized, with more overlap of when it started and finished.

Today the concept of human menstrual synchronization is generally known as the McClintock effect. But the idea that has continued to shape both her research and her reputation, the one that drives a still flourishing field of research, is that this mysterious synchrony, this reproductive networking, is caused by chemical messaging between women—the notion that humans, like so many other creatures, reach out to one another with chemical signals.

It has been harder than expected to single out specific signaling chemicals and trace their effects on our bodies and minds as precisely as entomologists have done for countless insect pheromones. But in the four decades since McClintock's discovery, scientists have charted the influence of chemical signaling across a spectrum of human behaviors. Not only do we synchronize our reproductive cycles, we can also recognize our kin, respond to others' stress, and react to their moods—such as fear or sadness or "not tonight, honey"—all by detecting chemicals they quietly secrete. As researchers learn more about this web of human interaction, they are helping to bridge an arbitrary dividing line between humans and the natural world.

Animal Kingdom Chemistry

The very intriguing idea of animals sharing invisible chemical cues has a long and illustrious history, at least as far as other species are concerned. The ancient Greeks talked enthusiastically of the possibility that female dogs in heat might produce some mysterious secretion capable of driving male dogs into a panting frenzy. Charles Darwin, pointing to several famously smelly species, proposed that chemical signals were part of the sexual selection process. Throughout the late nineteenth century the great French naturalist Jean-Henri Fabre puzzled over evidence that the siren call of chemistry could stir winged insects into determined flight.

Still, it was not until 1959 that the science really began to gain traction. In that year Adolf Butenandt, a Nobel laureate in chemistry, isolated and analyzed a compound that female silk moths release to attract males. Butenandt dissected the insects and painstakingly extracted the chemical from their microscopic secretion glands. He collected enough to crystallize it so that he could discern its molecular structure by x-ray crystallography. He called the compound bombykol, after the Latin name for the silk moth.

It was the first known pheromone, although the term did not yet exist. Shortly after, two of Butenandt's colleagues, the German biochemist Peter Karlson and the Swiss entomologist Martin Lüscher, coined that name from two Greek words: *pherein* (to transport) and *horman* (to stimulate). They defined a pheromone as a type of small molecule that carries chemical messages between individuals of the same species. The compounds must be active in very tiny amounts, potent below a conscious scent threshold. When released by one individual in a species and received by another, the two researchers wrote, they produce a measurable effect, "a specific reaction, for instance, a definite behavior or a developmental process."

Since then, an astonishing array of pheromones—the best-known and established class of chemical-signaling molecules exchanged by animals—have been found in insects, not just in silk moths but in bark beetles, cabbage looper moths, termites, leaf-cutter ants, aphids, and honeybees. According to a 2003 report from the National Academy of Sciences, entomologists "have now broken the code for the pheromone communication of more than 1,600 insects." And pheromones serve many more purposes than simply attracting mates: they elicit alarm, identify kin, alter mood, tweak relationships.

By the late 1980s pheromones had also been found to influence a wide spectrum of noninsect species, including lobsters, fish, algae, yeast, ciliates, bacteria, and more. As this new science of chemical communication grew—acquiring the more formal name of semiochemistry, from the Greek *semion* (meaning "signal")—scientists extended the search to mammals. Almost immediately they ran into resistance from their colleagues.

"In the 1970s and 1980s people would jump at you if you said 'mammalian pheromone,'" recalls Milos Novotny, director of the Institute for Pheromone Research at Indiana University. "They'd say, 'There's no such thing: mammals are not like insects. They're too evolved and complex to be spontaneously responding to something like a pheromone.'"

But by the mid-1980s Novotny had not only identified a pheromone in mice that regulated intermale aggression, he had synthesized it. Such compounds were also verified in rats, hamsters, rabbits, and squirrels. And as the list lengthened, it also became apparent that mammals' pheromones were very like—if not identical to—those found in insects. As an example, most researchers cite the stunning work of the late Oregon Health and Science University biochemist L.E.L. "Bets" Rasmussen, who showed in 1996 that a sex pheromone secreted by female Asian elephants is chemically identical to one used by more than one hundred species of moths for similar purposes of attraction.

McClintock had proposed a similar idea in 1971 in her pioneering paper on menstrual synchrony. "Perhaps," she wrote then, "at least one female pheromone affects the timing of other female menstrual cycles."

Odorous Landscape

McClintock, now sixty-three, is sitting in a small, sunny room occupied by filing cabinets, computers, racks of stoppered vials and tubes, and scent sticks—all contributing to a faint, slightly sweet chemical aroma—and a dark-haired graduate student named David Kern. ("All the other graduate students would climb over my dead body to get in this room," he says.) McClintock's lab is at the University of Chicago's Institute for Mind and Biology, of which she is a founding director. She wears a tweedy jacket over a bright patterned shirt, and she is thinking over a question: How far has the science of semiochemistry traveled since that day some four decades ago? The case for human chemical communication has been made, she says, and "our goal is to tackle identifying the chemical compounds. And then we can refine our understanding of what fundamental roles they play."

That task is anything but easy. Human body odor is estimated to derive from about 120 compounds. Most of these compounds occur in the water-rich solution produced by the sweat glands or are released from apocrine, or scent, glands in the oily shafts of hair follicles. The apocrine glands concentrate the most under the arms, around the nipples and in the genital regions.

It is a complicated landscape, made even more complicated by our use of what researchers refer to as exogenous compounds, such as soap, deodorants, and perfumes, as Johan Lundström of the Monell Chemical Senses Center in Philadelphia points out. And yet Lundström marvels at how adeptly our brains sort through this chemical tangle. Neuroimaging work done at his lab finds a 20 percent faster response to known human chemical signals compared with chemically similar molecules found elsewhere in the environment. "The brain always knows when it smells a body odor," Lundström says.

This capacity is already present in infancy. Numerous studies in humans have shown that, as is true in animals, mothers and infants are acutely attuned to each other's scent. This scent knowledge is so precise that babies even prefer the parts of clothes worn by their mother (and their mother only) touched by sweat compounds. The recognition, interestingly, is more acute in breast-fed infants than in those raised on baby formula.

"We're still just mapping the influential compounds from those that are not," Lundström says. "I don't think we're dealing with one single compound but rather a range of different ones that may be important at different times." Pheromones operate under the radar, he says, and they influence—but do not necessarily completely control—numerous behaviors. "If we compare these with social cues, they may be less important than the obvious ways we communicate," Lundström says. But, he adds, the ability probably aided survival as we evolved, keeping us more closely attuned to one another.

Psychologist Denise Chen of Rice University also argues that this kind of chemical alertness would have conferred an evolutionary advantage. In her research, she collects odor samples from individuals while they watch horror movies. Gauze pads are kept in viewers' armpits to collect sweat released during moments of fear. Later the pads are placed under volunteers' nostrils. For comparison, Chen has also collected sweat from people watching comedies or neutral films such as documentaries.

One of her early experiments found that participants could tell whether the sweat donor was fearful or happy at the time the sweat was produced. The subjects' guesses succeeded more often than they would by pure chance, especially for fear-induced sweat. Chen followed up with research showing that exposure to "fear sweat" seemed to intensify the alarm response—inclining participants to see fear in the faces of others. These exposures even enhanced cognitive performance: on word-association tests that included terms suggestive of danger, women smelling fear sweat outperformed those exposed to neutral sweat. "If you smell fear, you're faster at detecting fearful words," Chen explains.

In a recent study, she and Wen Zhou of the Chinese Academy of Sciences compared the responses of long-time couples with those of people in shorter-term relationships, Those results indicated —perhaps not surprisingly—that the longer couples are together, the better the partners are at interpreting the fear or happiness information apparently encoded in sweat. "What I hope that people will see in this is that understanding olfaction is important for us to understand ourselves," Chen says.

And evidence continues to accumulate that unconscious perception of scents influences a range of human behaviors, from cognitive to sexual. In January, for instance, a team of scientists at Israel's Weizmann Institute of Science in Rehovot, led by neurobiologist Noam Sobel, reported that men who sniffed drops of women's emotional tears felt suddenly less sexually interested, in comparison to those who smelled a saline solution. Sobel found a direct physical response to this apparent chemosignal: a small but measurable drop in the men's testosterone levels. The signal may have evolved to signify lower fertility, as during menstruation. More generally, the discovery may help explain the uniquely human behavior of crying.

Hard Science

A major goal now is to identify the key chemicals that convey signals surreptitiously and to learn much more about how the body detects and reacts to those signals. George Preti, a Monell chemist, has mapped out a research project that would include tracking these messengers by analyzing sweat and apocrine secretions and studies of hormone levels in those who sniff the chemicals. "We've yet to identify the precise signals that carry the information," Lundström agrees. "And if we want a solid standing for this work, that's what's needed next."

McClintock also sees this as a priority. In recent years she has focused on building a detailed portrait of one of the more potent known chemosignals, a steroid compound called androstadienone. She believes that this particular small molecule is potent enough to meet the requirements of being called a human pheromone: it is a small molecule that acts as a same-species chemical signal and influences physiology and behavior. Over the years, labs, including McClintock's and Lundström's, have found that this particular compound shows measurable effects on cognition and that it can alter levels of stress hormones such as cortisol and evoke changes in emotional response.

In one recent study McClintock and her colleague Suma Jacob of the University of Illinois at Chicago explored androstadienone's propensity to affect mood. They mixed a trace amount into the solvent propylene glycol and then masked any possible overt odor with oil of clove. They then exposed one study group to a solvent containing the compound and another group to a plain solvent. Subjects were asked to smell gauze pads containing one version; they were told only that they were participating in olfaction research. All the subjects went on to fill out a long and tedious questionnaire.

Overall, the subjects exposed to androstadienone remained far more cheerful throughout the fifteen- to twenty-minute test. A follow-up study repeated the same process but included brain imaging as well. The neuroimages showed that brain regions associated with attention, emotion, and visual processing were more active in those exposed to the chemosignaling compound. McClintock sees this as a classic pheromonal effect, the kind that she speculated about decades ago.

Even so, she and other researchers continue to carefully talk of "putative" pheromones. Humans are complicated, and any causal links between specific chemicals and changes in behavior are hard to demonstrate conclusively. Indeed, no one can say for certain yet what chemical or chemicals account for McClintock's original discovery, the synchronization of women's menstrual cycles. Even the phenomenon itself has proved somewhat elusive: it has been confirmed in numerous follow-up studies but contradicted by others, and it is still not accepted unanimously by the scientific community.

Much of the discussion centers on what exactly is being synchronized—perhaps timing of ovulation, perhaps length of cycle. A review of human data from the 1990s by the father-and-son team of Leonard and Aron Weller of Bar-Ilan University in Israel found that synchrony sometimes occurs and sometimes does not. "If it exists," Leonard Weller reported, "it is certainly not ubiquitous."

Although she still retains the assertiveness of her college days, McClintock agrees that the effect is subtler than she thought at first. But she also believes that the critics tend to miss the more important point: that evidence for chemical communication between humans has steadily accumulated since her study. And that it is not surprising that our chemical messaging is turning out to be as intricate as every other form of human communication.

ELIZABETH KOLBERT

Sleeping with the Enemy

FROM *The New Yorker*

THE MAX PLANCK INSTITUTE for Evolutionary Anthropology, in Leipzig, is a large, mostly glass building shaped a bit like a banana. The institute sits at the southern edge of the city, in a neighborhood that still very much bears the stamp of its East German past. If you walk down the street in one direction, you come to a block of Soviet-style apartment buildings; in the other, to a huge hall with a golden steeple, which used to be known as the Soviet Pavilion. (The pavilion is now empty.) In the lobby of the institute there's a cafeteria and an exhibit on great apes. A TV in the cafeteria plays a live feed of the orangutans at the Leipzig Zoo.

Svante Pääbo heads the institute's department of evolutionary genetics. He is tall and lanky, with a long face, a narrow chin, and bushy eyebrows, which he often raises to emphasize some sort of irony. Pääbo's office is dominated by a life-size model of a Neanderthal skeleton, propped up so that its feet dangle over the floor, and by a larger-than-life-size portrait that his graduate students presented to him on his fiftieth birthday. Each of the students painted a piece of the portrait, the overall effect of which is a surprisingly good likeness of Pääbo, but in mismatched colors that make it look as if he had a skin disease.

At any given moment, Pääbo has at least half a dozen research efforts in progress. When I visited him in May, he had one team analyzing DNA that had been obtained from a 40,000- or 50,000-year-old finger bone found in Siberia, and another trying to extract DNA from a cache of equally ancient bones from China.

A third team was slicing open the brains of mice that had been genetically engineered to produce a human protein.

In Pääbo's mind, at least, these research efforts all hang together. They are attempts to solve a single problem in evolutionary genetics, which might, rather dizzyingly, be posed as: What made us the sort of animal that could create a transgenic mouse?

The question of what defines the human has, of course, been kicking around since Socrates, and probably a lot longer. If it has yet to be satisfactorily resolved, then this, Pääbo suspects, is because it has never been properly framed. "The challenge is to address the questions that are answerable," he told me.

Pääbo's most ambitious project to date, for which he has assembled an international consortium to assist him, is an attempt to sequence the entire genome of the Neanderthal. The project is about halfway complete and has already yielded some unsettling results, including the news, announced by Pääbo last year, that modern humans, before doing in the Neanderthals, must have interbred with them.

Once the Neanderthal genome is complete, scientists will be able to lay it gene by gene—indeed, base by base—against the human, and see where they diverge. At that point, Pääbo believes, an answer to the age-old question will finally be at hand. Neanderthals were very closely related to modern humans—so closely that we shared our prehistoric beds with them—and yet clearly they were *not* humans. Somewhere among the genetic disparities must lie the mutation or, more probably, mutations that define us. Pääbo already has a team scanning the two genomes, drawing up lists of likely candidates.

"I want to know what changed in fully modern humans, compared with Neanderthals, that made a difference," he said. "What made it possible for us to build up these enormous societies, and spread around the globe, and develop the technology that I think no one can doubt is unique to humans. There has to be a genetic basis for that, and it is hiding somewhere in these lists."

Pääbo, who is now fifty-six, grew up in Stockholm. His mother, a chemist, was an Estonian refugee. For a time, she worked in the laboratory of a biochemist named Sune Bergström, who later won a Nobel Prize. Pääbo was the product of a lab affair between the two, and, although he knew who his father was, he wasn't supposed to discuss it. Bergström had a wife and another son; Pääbo's mother, meanwhile, never married. Every Saturday, Bergström would visit Pääbo and take him for a walk in the woods or somewhere else where he didn't think he'd be recognized.

"Officially, at home, he worked on Saturday," Pääbo told me. "It was really crazy. His wife knew. But they never talked about it. She never tried to call him at work on Saturdays." As a child, Pääbo wasn't particularly bothered by the whole arrangement; later, he occasionally threatened to knock on Bergström's door. "I would say, 'You have to tell your son—your other son—because he will find out sometime,'" he recalled. Bergström would promise to do this but never followed through. (As a result, Bergström's other son did not learn that Pääbo existed until shortly before Bergström's death, in 2004.)

From an early age, Pääbo was interested in old things. He discovered that around fallen trees it was sometimes possible to find bits of pottery made by prehistoric Swedes, and he filled his room with potsherds. When he was a teenager, his mother took him to visit the pyramids, and he was entranced. He enrolled at Uppsala University, planning to become an Egyptologist.

"I really wanted to discover mummies, like Indiana Jones," he said. Mostly, though, the coursework turned out to involve parsing hieroglyphics, and instead of finding it swashbuckling Pääbo thought it was boring. Inspired by his father, he switched first to medicine, then to cell biology.

In the early 1980s, Pääbo was doing doctoral research on viruses when he once again began fantasizing about mummies. At least as far as he could tell, no one had ever tried to obtain DNA from an ancient corpse. It occurred to him that if this was possible, then a whole new way of studying history would open up.

Suspecting that his dissertation adviser would find the idea silly (or worse), Pääbo conducted his mummy research in secret, at night. With the help of one of his former Egyptology professors, he managed to obtain some samples from the Egyptian Museum in what was then East Berlin. In 1984, he published his results in an obscure East German journal. He had, he wrote, been able to detect DNA in the cells of a mummified child who'd been dead for more than two thousand years. Among the questions that Pääbo thought mummy DNA could answer were what caused pharaonic dynasties to change and who Tutankhamen's mom was.

While Pääbo was preparing a version of his mummy paper for publication in English, a group of scientists from the University of California at Berkeley announced that they had succeeded in sequencing a snippet of DNA from a zebralike animal known as a quagga, which had been hunted to extinction in the 1880s. (The DNA came from a 140-year-old quagga hide preserved at the National History Museum in Mainz.) The leader of the team, Allan Wilson, was an eminent biochemist who had, among other things, come up with a way to study evolution using the concept of a "molecular clock." Pääbo sent Wilson the galleys of his mummy paper. Impressed, Wilson replied, asking if there was any space in Pääbo's lab; he might like to spend a sabbatical there. Pääbo had to write back that he could not offer Wilson space in his lab, because, regrettably, he didn't have a lab—or even, at that point, a PhD.

Pääbo's mummy paper became the cover article in *Nature.* It was also written up in the *New York Times,* which called his achievement "the most dramatic of a series of recent accomplishments using molecular biology." Pääbo's colleagues in Sweden, though, remained skeptical. They urged him to forget about shriveled corpses and stick to viruses.

"Everybody told me that it was really stupid to leave that important area for something which looked like a hobby of some sort," he said. Ignoring them, Pääbo moved to Berkeley to work for Wilson.

"He just kind of glided in," Mary-Claire King, who had also been a student of Wilson's and who is now a professor of genome sciences at the University of Washington, recalled. According to King, Pääbo and Wilson, who died in 1991, turned out to share much more than an interest in ancient DNA.

"Each of them thought of very big ideas," she told me. "And each of them was very good at translating those ideas into testable hypotheses. And then each of them was very good at developing the technology that's necessary to test the hypotheses. And to have all three of those capacities is really remarkable." Also, although "they were both very data-driven, neither was afraid to say outrageous things about their data, and neither was afraid to be wrong."

DNA is often compared to a text, a comparison that's apt as long as the definition of "text" encompasses writing that doesn't make sense. DNA consists of molecules known as nucleotides knit together in the shape of a ladder—the famous double helix. Each nucleotide contains one of four bases: adenine, thymine, guanine, and cytosine, which are designated by the letters A, T, G, and C, so that a stretch of the human genome might be represented as ACCTCCTCTAATGTCA. (This is an actual sequence, from chromosome 10; the comparable sequence in an elephant is ACCTCCCCTAATGTCA.) The human genome is 3 billion bases—or, really, base pairs—long. As far as can be determined, most of it is junk.

With the exception of red blood cells, every cell in an organism contains a complete copy of its DNA. It also contains many copies—from hundreds to thousands—of an abridged form of DNA known as mitochondrial DNA, or mtDNA. But as soon as the organism dies, the long chains of nucleotides begin to break down. Much of the damage is done in the first few hours, by enzymes inside the creature's own body. After a while, all that remains is snippets, and after a longer while—how long seems to depend on the conditions of decomposition—these snippets, too, disintegrate. "Maybe in the permafrost you could go back five hundred thousand years," Pääbo told me. "But it's certainly on this side of a million." Five hundred thousand years ago, the dinosaurs had been dead for more than 64 million years, so the whole *Jurassic Park* fantasy is, sadly, just that. On the other hand, 500,000 years ago modern humans did not yet exist.

When Pääbo arrived in California, he was still interested in finding a way to use genetics to study human history. He'd discovered, however, a big problem with trying to locate fragments of ancient Egyptian DNA: they look an awful lot like—indeed, identical to—fragments of contemporary human DNA. Thus a single microscopic particle of his own skin, or of someone else's, even some long-dead museum curator's, could nullify months of work.

"It became clear that human contamination was a huge problem," he explained. (Eventually, Pääbo concluded that the sequences he had obtained for his original mummy paper had probably been corrupted in this way.) As a sort of warmup exercise, he began working on extinct animals. He analyzed scraps of mtDNA from giant ground sloths, which disappeared about 12,000 years ago, and from mammoths, which vanished around the same time, and from Tasmanian tigers, which were hunted to extinction by the 1930s. He extracted mtDNA from moas, the giant flightless birds that populated New Zealand before the arrival of the Maori, and found that moas were more closely related to birds from Australia than to kiwis, the flightless birds that inhabit New Zealand today. "That was a blow to New Zealand self-esteem," he recalled. He also probed plenty of remains that yielded no usable DNA, including bones from the La Brea tar pits and fossilized insects preserved in amber. In the process of this work, Pääbo more or less invented the field of paleogenetics.

"Frankly, it was a problem that I wouldn't have tackled myself, because I thought it was too difficult," Maynard Olson, an emeritus professor at the University of Washington and one of the founders of the Human Genome Project, told me. "Pääbo brought very high standards to this area, and took the field of ancient DNA study from its *Jurassic Park* origins to a real science, which is a major accomplishment."

"There's nothing unique about most of science," Ed Green, a professor of biomolecular engineering at the University of California at Santa Cruz who works on the Neanderthal Genome Project, said. "If you don't do it, somebody else is going to do it a few months later. Svante is one of the rare people in science for whom that is not true. There wouldn't even be a field of ancient DNA as we know it without him."

"It's a nice rarity in science when people take not only unique but also productive paths," Craig Venter, who led a rival effort to the Human Genome Project, told me. "And Svante has clearly done both. I have immense respect for him and what he's done."

While Pääbo was living in California, he sometimes went to Germany to visit a woman who was attending graduate school at the University of Munich. "I had many relationships with men, but I also had girlfriends now and again," he told me. The relationship ended; shortly afterward, the University of Munich offered Pääbo an assistant professorship. With no pressing reason to move to Germany, he demurred. The offer was increased to a full professorship: "So then I said, 'Germany isn't that bad after all. I'll go there for a few years.'"

Pääbo was still in Munich several years later when he got a call from the Rhenish State Museum in Bonn. The museum houses the bones of the first Neanderthal to be identified as such, which was discovered in the summer of 1856. What did Pääbo think the odds were that he could extract usable DNA? He had no way of determining what kind of shape the bones were in until he dissolved them.

"I didn't know what to tell them, so I said, 'There's a five-percent chance that it works,'" he recalled. A few months later, he received a small chunk of the Neanderthal's right humerus.

The first Neanderthal was found in a limestone cave about forty-five miles north of Bonn, in an area known as the Neander Valley, or, in German, *das Neandertal*. Although the cave is gone—the limestone was long ago quarried into building blocks—the area is now a sort of Neanderthal theme park, with its own museum, hiking trails, and a garden planted with the kinds of shrubs that would have been encountered during an ice age. In the museum, Neanderthals are portrayed as kindly, if not particularly telegenic, humans. By the entrance to the building, there's a model of an elderly Neanderthal leaning on a stick. He is smiling benignantly and resembles an unkempt Yogi Berra. Next to him is one of the museum's most popular attractions—a booth called the Morphing-Station. For three euros, visitors to the station can get a normal profile shot of themselves and, facing that, a second shot that has been doctored. In the second, the chin recedes, the forehead slopes, and the back of the head bulges out. Kids love to see themselves—or, better yet, their siblings—morphed into Neanderthals. They find it screamingly funny.

When the first Neanderthal bones showed up in the Neander Valley, they were treated as rubbish (and almost certainly damaged in the process). The fragments—a skullcap, four arm bones, two thighbones, and part of a pelvis—were later salvaged by a local businessman, who, thinking they belonged to a cave bear, passed them on to a fossil collector. The fossil collector realized that he was dealing with something much stranger than a bear. He declared the remains to be traces of a "primitive member of our race."

As it happened, this was right around the time that Darwin published *On the Origin of Species*, and the fragments soon got caught up in the debate over the origin of humans. Opponents of evolution insisted that they belonged to an ordinary person. One theory held that it was a Cossack who had wandered into the region in the tumult following the Napoleonic Wars. The reason the bones looked odd—Neanderthal femurs are distinctly bowed—was that the Cossack had spent too long on his horse. Another attributed the remains to a man with rickets: the man had been in so much pain from his disease that he'd kept his forehead perpetually tensed—hence the protruding brow ridge. (What a man with rickets and in constant pain was doing climbing into a cave was never really explained.)

Over the next decades, bones resembling those from the Neander Valley—thicker than those of modern humans, with strangely shaped skulls—were discovered at several more sites, including two in Belgium and one in France. Meanwhile, a skull that had been unearthed years earlier in Gibraltar was shown to look much like the one from Germany. Clearly, all these remains could not be explained by stories of disoriented Cossacks or rachitic spelunkers. But evolutionists, too, were perplexed by them. Neanderthals had very large skulls—larger, on average, than people today. This made it hard to fit them into an account of evolution that started with small-brained apes and led, through progressively bigger brains, up to humans. In *The Descent of Man,* which appeared in 1871, Darwin mentioned Neanderthals only in passing. "It must be admitted that some skulls of very high antiquity, such as the famous one of Neanderthal, are well developed and capacious," he noted.

In 1908 a nearly complete Neanderthal skeleton was discovered in a cave near La Chapelle-aux-Saints, in southern France. The skeleton was sent to a paleontologist named Marcellin Boule, at Paris's National Museum of Natural History. In a series of monographs, Boule invented what might be called the cartoon version of the Neanderthals—bent-kneed, hunched over, and brutish. Neanderthal bones, Boule wrote, displayed a "distinctly simian arrangement," while the shape of their skulls indicated "the predominance of functions of a purely vegetative or bestial kind." Boule's conclusions were studied and then echoed by many of his contemporaries; the British anthropologist Sir Grafton Elliot Smith, for instance, described Neanderthals as walking with "a half-stooping slouch" upon "legs of a peculiarly ungraceful form." (Smith also claimed that Neanderthals' "unattractiveness" was "further emphasized by a shaggy covering of hair over most of the body," although there was—and still is—no clear evidence that they were hairy.)

In the 1950s, a pair of anatomists, Williams Straus and Alexander Cave, decided to reexamine the skeleton from La Chapelle.

What Boule had taken for the Neanderthal's natural posture, Straus and Cave determined, was probably a function of arthritis. Neanderthals did not walk with a slouch or with bent knees. Indeed, given a shave and a new suit, the pair wrote, a Neanderthal probably would attract no more attention on a New York City subway "than some of its other denizens." More recent scholarship has tended to support the idea that Neanderthals, if not quite up to negotiating the IRT, certainly walked upright, with a gait we would recognize more or less as our own. The version of Neanderthals offered by the Neanderthal Museum—another cartoon—is imbued with cheerful dignity. Neanderthals are presented as living in tepees, wearing what look like leather yoga pants, and gazing contemplatively over the frozen landscape. "Neanderthal man was not some prehistoric Rambo," one of the display tags admonishes. "He was an intelligent individual."

Pääbo announced his plan to sequence the entire Neanderthal genome in July 2006, just in time for the hundred-and-fiftieth anniversary of the Neanderthal's discovery. The announcement was made together with an American company, 454 Life Sciences, which had developed a so-called high-throughput sequencing machine that, with the help of tiny resin spheres, could replicate tens of thousands of DNA snippets at a time. Both inside and outside the genetics profession, the plan was viewed as wildly ambitious, and the project made international news. "A Study with a Lot of Balls," the headline in *The Economist* declared.

By this point, a complete version of the human genome had been published. So, too, had versions of the chimpanzee, mouse, and rat genomes. But humans, chimps, mice, and rats are all living organisms, while Neanderthals have been extinct for 30,000 years. The first hurdle was simply finding enough Neanderthal DNA to sequence. The chunk of the original Neanderthal that Pääbo had received had yielded shreds of genetic information, but nowhere near the quantities needed to assemble—or reassemble—an entire genome. So Pääbo was placing his hopes on another set of bones, from Croatia. (The Croatian bones turned out to have belonged to three individuals, all of them women; the original Neanderthal was probably a man.)

Toward the end of 2006, Pääbo and his team reported that, using a piece of Croatian bone, they had succeeded in sequencing a million base pairs of the Neanderthal genome. (Just like the human genome, the full Neanderthal genome consists of roughly 3 billion base pairs.) Extrapolating from this, they estimated that to complete the project would take roughly two years and six thousand "runs" on a 454 Life Sciences machine. But later analysis revealed that the million base pairs had probably been contaminated by human DNA, a finding that led some geneticists to question whether Pääbo had rushed to publish results that he should have known were wrong. Meanwhile, subsequent bones yielded a much lower proportion of Neanderthal DNA and a much higher percentage of microbial DNA. (Something like 80 percent of the DNA that has been sequenced for the Neanderthal Genome Project belongs to microorganisms and, as far as the project is concerned, is useless.) This meant the initial estimates of the labor involved in finishing the genome were probably far too low. "There were times when one despaired," Pääbo told me. No sooner would one problem be resolved than another materialized. "It was an emotional roller coaster," Ed Green, the biomolecular engineer from Santa Cruz, recalled.

About two years into the project, a new puzzle arose. Pääbo had assembled an international team to help analyze the data the sequencing machines were generating—essentially, long lists of A's, T's, G's, and C's. Sifting through the data, one of the members of this team, David Reich, a geneticist at Harvard Medical School, noticed something odd. The Neanderthal sequences, as expected, were very similar to human sequences. But they were more similar to some humans than to others. Specifically, Europeans and Asians shared more DNA with Neanderthals than did Africans. "We tried to make this result go away," Reich told me. "We thought, This must be wrong."

For the past twenty-five years or so, the study of human evolution has been dominated by the theory known in the popular press as "Out of Africa" and in academic circles as the "recent single-origin" or "replacement" hypothesis. This theory holds that all modern humans are descended from a small population that lived in Africa roughly 200,000 years ago. (Not long before he died, Pääbo's adviser Allan Wilson developed one of the key lines of evidence for the theory, based on a comparison of mitochondrial DNA from contemporary humans.) Around 120,000 years ago, a subset of the population migrated into the Middle East, and by 50,000 years ago a further subset pushed into Eurasia. As they moved north and east, modern humans encountered Neanderthals and other so-called archaic humans who already inhabited those regions. The modern humans "replaced" the archaic humans, which is a nice way of saying they drove them into extinction. This model of migration and "replacement" implies that the relationship between Neanderthals and humans should be the same for all people alive today, regardless of where they come from.

Many members of Pääbo's team suspected another case of contamination. At various points, the samples had been handled by Europeans; perhaps their DNA had gotten mixed in with the Neanderthals'. Several tests were run to assess this possibility. The results were all negative. "We kept seeing this pattern, and the more data we got, the more statistically overwhelming it became," Reich told me. Gradually the other team members started to come around. In a paper published in *Science* in May 2010, they introduced what Pääbo has come to refer to as the "leaky replacement" hypothesis. (The paper was later voted the journal's outstanding article of the year, and the team received a $25,000 prize.) Before modern humans "replaced" the Neanderthals, they had sex with them. The liaisons produced children, who helped to people Europe, Asia, and the New World.

The leaky-replacement hypothesis—assuming for the moment that it is correct—provides further evidence of the closeness of Neanderthals to modern humans. Not only did the two interbreed; the resulting hybrid offspring were functional enough to be integrated into human society. Some of these hybrids survived to have kids of their own, who, in turn, had kids, and so on to the present day. Even now, at least 30,000 years after the fact, the signal is discernible: all non-Africans, from the New Guineans to the French to the Han Chinese, carry somewhere between 1 and 4 percent Neanderthal DNA.

One of Pääbo's favorite words in English is "cool." When he finally came around to the idea that Neanderthals bequeathed some of their genes to modern humans, he told me, "I thought it was very cool. It means that they are not totally extinct—that they live on a little bit in us."

The Leipzig Zoo lies on the opposite side of the city from the Institute for Evolutionary Anthropology, but the institute has its own lab building on the grounds, as well as specially designed testing rooms inside the ape house, which is known as Pongoland. Since none of our very closest relatives survive (except as little bits in us), researchers have to rely on our next closest kin, chimpanzees and bonobos, and our somewhat more distant cousins, gorillas and orangutans, for live experiments. (The same or at least analogous experiments are usually also performed on small children, to see how they compare.) One morning I went to the zoo, hoping to watch an experiment in progress. That day a BBC crew was also visiting Pongoland, to film a program on animal intelligence, and when I arrived at the ape house I found it strewn with camera cases marked ANIMAL EINSTEINS.

For the benefit of the cameras, a researcher named Héctor Marín Manrique was preparing to reenact a series of experiments he'd performed earlier in a more purely scientific spirit. A female orangutan named Dokana was led into one of the testing rooms. Like most orangutans, she had copper-colored fur and a world-weary expression. In the first experiment, which involved red juice and skinny tubes of plastic, Dokana showed that she could distinguish a functional drinking straw from a nonfunctional one. In the second, which involved more red juice and more plastic, she showed that she understood the *idea* of a straw by extracting a rod from a length of piping and using the pipe to drink through. Finally, in a Mensa-level show of pongid ingenuity, Dokana managed to get at a peanut that Manrique had placed at the bottom of a long plastic cylinder. (The cylinder was fixed to the wall, so it couldn't be knocked over.) She fist-walked over to her drinking water, took some water in her mouth, fist-walked back, and spat into the cylinder. She repeated the process until the peanut floated within reach. Later, I saw this experiment restaged with some five-year-old children, using little plastic containers of candy in place of peanuts. Even though a full watering can had been left conspicuously nearby, only one of the kids—a girl—managed to work her way to the floating option, and this was after a great deal of prompting. ("How would water help me?" one of the boys asked, just before giving up.)

One way to try to answer the question "What makes us human?" is to ask "What makes us different from apes?" or, to be more precise, from nonhuman apes since, of course, humans *are* apes. As just about every human by now knows—and as the experiments with Dokana once again confirm—nonhuman apes are extremely clever. They're capable of making inferences, of solving complex puzzles, and of understanding what others are (and are not) likely to know. When researchers from Leipzig performed a battery of tests on chimpanzees, orangutans, and two-and-a-half-year-old children, they found that the chimps, the orangutans, and the kids performed comparably on a wide range of tasks that involved understanding of the physical world. For example, if an experimenter placed a reward inside one of three cups and then moved the cups around, the apes found the goody just as often as the kids—indeed, in the case of chimps, more often. The apes seemed to grasp quantity as well as the kids did—they consistently chose the dish containing more treats, even when the choice involved using what might loosely be called math—and also seemed to have just as good a grasp of causality. (The apes, for instance, understood that a cup that rattled when shaken was more likely to contain food than one that did not.) And they were equally skillful at manipulating simple tools.

Where the kids routinely outscored the apes was in tasks that involved reading social cues. When the children were given a hint about where to find a reward—someone pointing to or looking at the right container—they took it. The apes either didn't understand that they were being offered help or couldn't follow the cue. Similarly, when the children were shown how to obtain a reward, by, say, ripping open a box, they had no trouble grasping the point and imitating the behavior. The apes, once again, were flummoxed. Admittedly, the kids had a big advantage in the social realm, since the experimenters belonged to their own species. But in general, apes seem to lack the impulse toward collective problem solving that's so central to human society.

"Chimps do a lot of incredibly smart things," Michael Tomasello, who heads up the institute's department of developmental and comparative psychology, told me. "But the main difference we've seen is 'putting our heads together.' If you were at the zoo today, you would never have seen two chimps carry something heavy together. They don't have this kind of collaborative project."

Pääbo usually works late, and most nights he has dinner at the institute, where the cafeteria stays open until seven P.M. One evening, though, he offered to knock off early and show me around downtown Leipzig. We visited the church where Bach is buried and ended up at Auerbachs Keller, the bar to which Mephistopheles brings Faust in the fifth scene of Goethe's play. (The bar was supposedly Goethe's favorite hangout when he was a university student.) Pääbo's wife, Linda Vigilant, an American primatologist who also works at the institute, joined us. Pääbo and Vigilant first met in the 1980s in Berkeley, but they didn't get together until both moved to Leipzig, in the late nineties. (Vigilant was then married to another geneticist, who works at the institute too.) Pääbo and Vigilant have a six-year-old son, and Vigilant has two older sons from her previous marriage.

I had been to the zoo, and I asked Pääbo about a hypothetical experiment. If he had the opportunity to subject Neanderthals to the sorts of tests I'd seen in Pongoland, what would he do? Did he think he'd be able to talk to them? He sat back in his chair and folded his arms across his chest.

"One is so tempted to speculate," he said. "So I try to resist it by refusing questions such as 'Do I think they would have spoken?' Because, honestly, I don't know, and in some sense you can speculate with just as much justification as I can."

By now, scores of Neanderthal sites have been excavated, from western Spain to central Russia and from Israel to Wales. They give lots of hints about what Neanderthals were like, at least for those inclined to speculate. Neanderthals were extremely tough—this is attested to by the thickness of their bones—and probably capable of beating modern humans to a pulp. They were adept at making stone tools, though they seem to have spent tens of thousands of years making the same tools over and over, with only marginal variation. At least on some occasions, they buried their dead. Also on some occasions, they appear to have killed and eaten each other. Wear on their incisors suggests that they spent a lot of time grasping animal skins with their teeth, which in turn suggests that they processed hides into some sort of leather. Neanderthal skeletons very often show evidence of disease or disfigurement. The original Neanderthal, from Mettmann, for example, seems to have suffered and recovered from two serious injuries, one to his head and the other to his left arm. The Neanderthal whose nearly complete skeleton was found in La Chapelle endured, in addition to arthritis, a broken rib and kneecap. Both individuals survived into their fifties, which indicates that Neanderthals had the capacity for collective action, or, if you prefer, empathy. They must—at least sometimes—have cared for their wounded.

From the archaeological record, it's inferred that Neanderthals evolved in Europe or western Asia and spread out from there, stopping when they reached water or some other significant obstacle. (During the ice ages, sea levels were a lot lower than they are now, so there was no English Channel to cross.) This is one of the most basic ways modern humans differ from Neanderthals and, in Pääbo's view, also one of the most intriguing. By about 45,000 years ago, modern humans had already reached Australia, a journey that, even mid–ice age, meant crossing open water. Archaic humans like *Homo erectus* "spread like many other mammals in the Old World," Pääbo told me. "They never came to Madagascar, never to Australia. Neither did Neanderthals. It's only fully modern humans who start this thing of venturing out on the ocean where you don't see land. Part of that is technology, of course; you have to have ships to do it. But there is also, I like to think or say, some madness there. You know? How many people must have sailed out and vanished on the Pacific before you found Easter Island? I mean, it's ridiculous. And why do you do that? Is it for the glory? For immortality? For curiosity? And now we go to Mars. We never stop." If the defining characteristic of modern humans is this sort of Faustian restlessness, then, by Pääbo's account, there must be some sort of Faustian gene. Several times he told me that he thought it should be possible to identify the basis for this "madness" by comparing Neanderthal and human DNA.

"If we one day will know that some freak mutation made the human insanity and exploration thing possible, it will be amazing to think that it was this little inversion on this chromosome that made all this happen and changed the whole ecosystem of the planet and made us dominate everything," he said at one point. At another, he said, "We are crazy in some way. What drives it? That I would really like to understand. That would be really, really cool to know."

According to the most recent estimates, Neanderthals and modern humans share a common ancestor who lived about 400,000 years ago. (It is unclear who that ancestor was, though one possibility is the somewhat shadowy hominid known, after a jawbone found near Heidelberg, as *Homo heidelbergensis*.) The common ancestor of chimps and humans, by contrast, lived some 5 million to 7 million years ago. This means that Neanderthals and humans had less than one-tenth the time to accumulate genetic differences.

Mapping these differences is, in principle, pretty straightforward—no harder, say, than comparing rival editions of *Hamlet*. In practice, it's quite a bit more complicated. To begin with, there's really no such thing as *the* human genome; everyone has his or her own genome, and they vary substantially—between you and the person sitting next to you on the subway, the differences are likely to amount to some 3 million base pairs. Some of these variations correspond to observable physiological differences—the color of your eyes, say, or your likelihood of developing heart disease—and some have no known significance. To a first approximation, a human and a Neanderthal chosen at random would also vary by 3 million base pairs. The trick is ascertaining which of these millions of variations divide us from them. Pääbo estimates that when the Neanderthal Genome Project is completed, the list of base-pair changes that are at once unique to humans and shared by all humans will number around 100,000. Somewhere in this long list will lie the change—or changes—that made us human to begin with. Identifying these key mutations is where the transgenic mice come in.

From an experimental viewpoint, the best way to test whether any particular change is significant would be to produce a human with the Neanderthal version of the sequence. This would involve manipulating a human stem cell, implanting the genetically modified embryo into a surrogate mother, and then watching the resulting child grow up. For obvious reasons, such *Island of Dr. Moreau*–like research on humans is not permitted, nor is it necessarily even possible. For similar reasons, such experimentation isn't allowed on chimpanzees. But it is allowed on mice. Dozens of strains of mice have been altered to carry humanized DNA sequences, and new ones are being created all the time, more or less to order.

Several years ago, Pääbo and a colleague, Wolfgang Enard, became interested in a gene known as FOXP2, which in humans is associated with language. (People who have a faulty copy of the gene—an extremely rare occurrence—are capable of speech, but what they say is, to strangers, mostly incomprehensible.) Pääbo and Enard had some mice bred with a humanized version of the gene and then studied them from just about every possible angle.

The altered mice, it turned out, squeaked at a lower pitch than their un-humanized peers. They also exhibited measurable differences in neural development. (While I was in Leipzig, I watched a graduate student cut the heads off some of the altered mice and then slice up their brains, like radishes.) The Neanderthals' FOXP2 gene, it turns out, is in almost all ways identical to humans', but there is one suggestive base-pair difference. When this difference was discovered, it prompted Pääbo to order up a new round of transgenic mice, which, at the time of my visit, had just been born and were being raised under sterile conditions in the basement.

Genes that seem to play a role in speech are obvious places to look for human-specific changes. But one of the main points of sequencing the Neanderthal genome is that the most obvious places to look may not be the right ones.

"The great advantage with genomics in this form is that it's unbiased," Pääbo told me. "If you go after candidate genes, you're inherently saying what you think the most important thing is. Language, many people would say. But perhaps we will be surprised —perhaps it's something else that was really crucial." Recently, Pääbo has become interested in a gene known as RUNX2, which is involved in bone formation. When members of his team analyzed the human and Neanderthal genomes mathematically, RUNX2 emerged as a place where significant changes in the human lineage seem to have occurred. People who have faulty copies of the RUNX2 gene often develop a condition, known as cleidocranial dysplasia, whose symptoms include such Neanderthal-like features as a flared rib cage. Two genes that have been implicated in autism, CADPS2 and AUTS2, also appear to have changed substantially between Neanderthals and humans. This is interesting because one of the symptoms of autism is an inability to read social cues.

One afternoon, when I wandered into his office, Pääbo showed me a photograph of a skullcap that had recently been discovered by an amateur collector about half an hour from Leipzig. From the photograph, which had been e-mailed to him, Pääbo had decided that the skullcap could be quite ancient—from an early Neanderthal or even a *Homo heidelbergensis*. He'd also decided that he had to have it. The skullcap had been found at a quarry in a pool of water—perhaps, he theorized, these conditions had preserved it, so that if he got to it soon, he'd be able to extract some DNA. But the skull had already been promised to a professor of anthropology in Mainz. How could he persuade the professor to give him enough bone to test?

Pääbo called everyone he knew who he thought might know the professor. He had his secretary contact the professor's secretary to get the professor's private cell-phone number, and joked—or maybe only half joked—that he'd be willing to sleep with the professor if that would help. The frenzy of phoning back and forth across Germany lasted for more than an hour and a half, until Pääbo finally talked to one of the researchers in his own lab. The researcher had seen the actual skullcap and concluded that it probably wasn't very old at all. Pääbo immediately lost interest in it.

With old bones, you never really know what you're going to get. A few years ago, Pääbo managed to get hold of a bit of tooth from one of the so-called "hobbit" skeletons found on the island of Flores, in Indonesia. (The "hobbits," who were discovered in 2004, are generally believed to have been diminutive archaic humans—*Homo floresiensis*—though some scientists have argued that they were just modern humans who suffered from microcephaly.) The tooth, which was about 17,000 years old, yielded no DNA.

Then, about a year and a half ago, Pääbo obtained a fragment of finger bone that had been unearthed in a cave in southern Siberia along with a weird, vaguely human-looking molar. The finger bone—about the size of a pencil eraser—was believed to be more than 40,000 years old. Pääbo assumed that it came either from a modern human or from a Neanderthal. If it proved to be the latter, then the site would be the farthest east that Neanderthal remains had been found.

In contrast to the hobbit tooth, the finger fragment yielded astonishingly large amounts of DNA. When the analysis of the first bits was completed, Pääbo happened to be in the United States. He called his office, and one of his colleagues said to him, "Are you sitting down?" The DNA showed that the digit could not have belonged to a Neanderthal or to a modern human. Instead, its owner must have been part of some entirely different and previously unsuspected type of hominid. In a paper published in December 2010 in *Nature,* Pääbo and his team dubbed this group the Denisovans, after the Denisova Cave, where the bone had been found. "Giving Accepted Prehistoric History the Finger," ran the headline on the story in the Sydney *Morning Herald.* Amazingly —or perhaps, by now, predictably—modern humans must have interbred with Denisovans, too, because contemporary New Guineans carry up to 6 percent Denisovan DNA. (Why this is true of New Guineans but not native Siberians or Asians is unclear, but presumably has to do with patterns of human migration.)

It has been understood for a long time that modern humans and Neanderthals were contemporaries. The discovery of the hobbits and now the Denisovans shows that humans shared the planet with at least two additional creatures like ourselves. And it seems likely that as DNA from more ancient remains is analyzed, still other human relatives will be found; as Chris Stringer, a prominent British paleoanthropologist, told me, "I'm sure we've got more surprises to come."

"If these other forms of humans had survived two thousand generations more, which is not so much, then how would that have influenced our view of the living world?" Pääbo said, once the excitement over the skullcap had passed and we were sitting over coffee. "We now make this very clear distinction between humans and animals. But it might not be as clear. That is sort of an interesting thing to philosophize about." It's also interesting to think about why we're the ones who survived.

Over the decades, many theories have been offered to explain what caused the demise of the Neanderthals, ranging from climate change to simple bad luck. In recent years, though, it's become increasingly clear that, as Pääbo put it to me, "their bad luck was us." Again and again, the archaeological evidence in Europe indicates, once modern humans showed up in a region where Neanderthals were living, the Neanderthals in that region vanished. Perhaps the Neanderthals were actively pursued, or perhaps they were just outcompeted. The Neanderthals' "bad luck" is presumably the same misfortune that the hobbits and the Denisovans encountered, and similar to the tragedy suffered by the giant marsupials that once browsed Australia, and the varied megafauna that used to inhabit North America, and the moas that lived in New Zealand. And it is precisely the same bad luck that has brought so many species

—including every one of the great apes—to the edge of oblivion today.

"To me, the mystery is not the extinction of the Neanderthals," Jean-Jacques Hublin, the director of the Institute for Evolutionary Anthropology's department of human evolution, told me. "To me the mystery is what makes modern humans such a successful group that they have been replacing not just the Neanderthals but *everything*. We don't have much evidence that the Neanderthals or other archaic humans ever led to an extinction of a species of mammal or anything else. For modern humans, there are hundreds of examples, and we do it very well."

One of the largest assemblages of Neanderthal bones ever found—remains from seven individuals—was discovered about a century ago at a spot known as La Ferrassie, in southwestern France. La Ferrassie is in the Dordogne, not far from La Chapelle and within half an hour's drive of dozens of other important archaeological sites, including the painted caves at Lascaux. Over the summer, a team that included one of Pääbo's colleagues was excavating at La Ferrassie, and I decided to go down and have a look. I arrived at the dig's headquarters—a converted tobacco barn—just in time for a dinner of boeuf bourguignonne, which was served on makeshift tables in the backyard.

The next day, I drove out to La Ferrassie with some of the team's archaeologists. The site lies in a sleepy rural area, right by the side of the road. Many thousands of years ago, La Ferrassie was a huge limestone cave, but one of the walls has since fallen in, and now it is open on two sides. A massive ledge of rock juts out about twenty feet off the ground, like half of a vaulted ceiling. The site is ringed by a wire fence and hung with tarps, which give it the aspect of a crime scene.

The day was hot and dusty. Half a dozen students crouched in a long trench, picking at the dirt with trowels. Along the side of the trench, I could see bits of bone sticking out from the reddish soil. The bones toward the bottom, I was told, had been tossed there by Neanderthals. The bones near the top were the leavings of modern humans, who occupied La Ferrassie once the Neanderthals were gone. The Neanderthal skeletons from the site had long since been removed, but there was still hope that some stray bit, like a tooth, might be found. Each bone fragment that was unearthed, along with every flake of flint and anything else that might even remotely be of interest, was set aside to be taken back to the headquarters to be sorted and tagged.

After watching the students chip away for a while, I retreated to the shade. I tried to imagine what life had been like for the Neanderthals at La Ferrassie. Though the area is now wooded, then it would have been tundra. There would have been elk roaming the valley, and reindeer and wild cattle and mammoths. Beyond these stray facts, not much came to me. I put the question to the archaeologists I had driven out with.

"It was cold," Shannon McPherron, of the Max Planck Institute, volunteered.

"And smelly," Dennis Sandgathe, of Canada's Simon Fraser University, said.

"Probably hungry," Harold Dibble, of the University of Pennsylvania, added.

"No one would have been very old," Sandgathe said.

Later on, back at the barn, I picked through the bits and pieces that had been dug up over the past few days. There were hundreds of fragments of animal bone, each of which had been cleaned and numbered and placed in its own little plastic bag, and hundreds of flakes of flint. Most of the flakes were probably the detritus of toolmaking—the Stone Age equivalent of wood shavings—but some, I learned, were the tools themselves. Once I was shown what to look for, I could see the beveled edges that the Neanderthals had crafted. One tool in particular stood out: a palm-size flint shaped like a teardrop. In archaeological terms, it was a hand axe, though it probably was not used as an axe in the contemporary sense of the word. It had been found near the bottom of the trench, so it was estimated to be about 70,000 years old. I took it out of its plastic bag and turned it over. It was almost perfectly symmetrical and—to a human eye, at least—quite beautiful. I said that I thought the Neanderthal who had fashioned it must have had a keen sense of design. McPherron objected.

"We know the end of the story," he told me. "We know what modern culture looks like, and so then what we do is we want to explain how we got here. And there's a tendency to overinterpret the past by projecting the present onto it. So when you see a beautiful hand axe and you say, 'Look at the craftsmanship on this; it's virtually an object of art,' that's your perspective today. But you can't assume what you're trying to prove."

Among the hundreds of thousands of Neanderthal artifacts that have been unearthed, almost none represent unambiguous attempts at art or adornment, and those that have been interpreted this way—for instance, ivory pendants discovered in a cave in central France—are the subject of endless, often abstruse disputes. (Many archaeologists believe that the pendants were created by Neanderthals who had come into contact with modern humans and were trying to imitate them, but, relying on the most recent dating techniques, some argue that the pendants were, in fact, created by modern humans.) This paucity has led some to propose that Neanderthals were not capable of art or—what amounts to much the same thing—not interested in it. They simply did not possess what, genomically speaking, might be called the aesthetic mutation.

On my last day in the Dordogne, I decided to visit a nearby human site known for its extraordinary images. The site, Grotte des Combarelles, is a long, very narrow cave that zigzags through a limestone cliff. Hundreds of feet in, the walls of the cave are covered with engravings—a mammoth curling its trunk, a wild horse lifting its head, a reindeer leaning forward, apparently to drink. In very recent times, the floor of the Grotte des Combarelles has been dug out, so that a person can walk in it, and the tunnel is dimly lit by electric lights. But when the etchings were originally created, some 12,000 or 13,000 years ago, the only way to gain access to the site would have been to crawl, and the only way to see in the absolute dark would have been to carry fire. As I crept along through the gloom, past engravings of wisent and aurochs and woolly rhinos, it occurred to me that I really had no clue what would drive someone to wriggle through a pitch-black tunnel to cover the walls with images that only another, similarly driven soul would see. Yet it also struck me that so much of what is distinctively human was here on display—creativity, daring, "madness." And then there were the animals pictured on the walls—the aurochs and mammoths and rhinos. These were the beasts that Paleolithic Europeans had hunted and then, one by one, as with the Neanderthals, obliterated.

MICHAEL ROBERTS

The Touchy-Feely (but Totally Scientific!) Methods of Wallace J. Nichols

FROM *Outside*

THE PHILIPPINE CORAL REEF tank inside the California Academy of Sciences in San Francisco is 25 feet deep and holds 212,000 gallons of water, making it one of the largest exhibits of living coral anywhere in the world. It is the centerpiece of the academy's Steinhart Aquarium and hosts hundreds of coral species, a couple thousand colorful fish, plus sharks, stingrays, and numerous smaller creatures, like sea anemones and snails. There are five windows affording looks inside, the biggest of which, at 16½ feet tall and almost 30 feet wide, makes a sweeping arc in front of a dimly lit standing area backed by several rows of benches. It was designed to offer visitors a panoramic, theaterlike view of life in the tank and is among the museum's most popular attractions. It's Wallace J. Nichols's favorite spot in the building.

Nichols, forty-four, is a biologist and research associate at the academy who made a name for himself in the mid-1990s when he tracked a loggerhead turtle that swam from Baja, Mexico, to Japan, the first time anyone had recorded an animal swimming an entire ocean. He has done fieldwork in waters around the globe and spends most of his waking hours thinking and talking about the ocean, but when he's in front of that big window at the aquarium, he doesn't watch the fish. He watches the people.

"Whether it's a ninety-two-year-old or a two-year-old, when they come into that blue space, something happens," Nichols says. They grow quiet and calm, but there's more to it than that. When couples walk in, they frequently start holding hands. He says that if you ask people here what they're feeling, they'll struggle for words. Nichols finds this fascinating. He also believes that if we can understand what really happens to us in the presence of the ocean—which brain processes underlie our emotional reactions—it could bring about a radical shift in conservation efforts. If we learn precisely *why* we love the ocean, his thinking goes, we'll have an immensely powerful new tool to protect it.

Not surprisingly, this theory can strike many of his peers as soft. "'You must be from California.' That's the first response," Nichols says. (He lives north of Santa Cruz, though he was raised in New Jersey.) But Nichols's credibility as a scientist, along with his charm and passion, have enabled him to rally excitement for his ideas among a diverse constituency of researchers and activists. In the past couple of years, he's become a sought-after speaker, giving dozens of presentations at a wide mix of venues, from TEDx to adventure-travel trade shows to environmental symposiums. His pitch: more data on rising sea temperatures or plastic pollution or disappearing creatures won't do anything for ocean conservation. Instead, we need to study our own minds.

Nichols envisions cognitive neuroscientists constructing detailed models of brain activity for experiences like sitting on a beach, then using their findings to drive public policy. "If I walk into a meeting of a coastal zoning commission and say, 'I think people listening to the ocean is good for them,' you'd see all the eyeballs in the room rolling," says Nichols. "But if I walk in and say, 'This is my friend the Stanford neuroscientist, and his research using brain scans shows that sitting by the ocean has the same calming effects as meditation on reducing stress,' suddenly access to the coast becomes a public health issue."

It's a viable fantasy that derives from the fact that Nichols himself isn't a neuroscientist. Unable to test his hypotheses, he's launched a campaign to create a new field of study he calls neuroconservation. His hope is to inspire cognitive scientists to examine these fundamental questions. As he sees it, it's a ripe invitation: Who wants to know what happens when our most complex organ meets the planet's largest feature?

"My role is to be the catalyst and cheerleader," he says. "But the question is, How do you turn this big idea into a movement?"

The first time I met Nichols, he gave me a blue marble. It was sort of awkward. "Hold it at arm's length," he said. "That's what the Earth looks like from a million miles away—a water planet. Now hold it up to your eye and look at the sun. If water were inside, it would contain virtually every element. Now think of someone who's doing good work for the ocean. Hold it to your heart: think of how it would feel to you and to them if you randomly gave them this marble as a way of saying thank you."

We were seated outside the Academy of Sciences on a late-winter afternoon. Nichols, who goes by J., was dressed in a casual button-down blue shirt, brown cords, and leather boots and wearing a perfectly manicured salt-and-pepper stubble beard. He looked directly into my eyes, speaking in a slow, even canter that was mildly hypnotic, the vestige of a stutter he overcame twenty-five years ago by forcing himself to make turtle presentations to school groups.

The marble shtick may have made me uncomfortable, but the last line stuck with me; I imagined myself giving the marble to an old friend. Turns out I'm not the only one to fall under this spell. Nichols tried it out for the first time in 2009, during a talk at the New England Aquarium in Boston, and the audience response was overwhelming. He figured he was on to something, so he set up a simple web site, BlueMarbles.org, and decided "to try and see how big we could make this thing with no budget or strategic plan." Nichols now estimates that there are as many as a million of his blue marbles in circulation around the planet. They have made it into the hands of Jane Goodall, Harrison Ford, James Cameron, E. O. Wilson, and the four-time Iditarod champion Lance Mackey, who carried one during this year's race.

Nichols's success at reaching large numbers of people on an emotional level both underscores the premise of his theories and makes it harder to dismiss him as a left-coast flake. Several times over the past six months, I watched him captivate audiences with a clever Trojan horse narrative: I'm a scientist, but—surprise!—I want to talk about how much we all *love* the ocean. During one lecture at Stanford, he implored graduate students to remember that as conservationists, "We have the power of happiness on our side."

For environmentalists struggling to find a message with staying power, Nichols's feel-good approach offers a compelling alternative to the usual tactic of scaring people into action with bad news about extinctions or global warming. "Hell, we've tried everything else," says the Nature Conservancy scientist M. Sanjayan. "We've tried to price nature. We've tried to stand and protest. We've tried every way we know to get people to see what we've seen, and we've been failing."

Nichols blames these failures on the detached way scientists gather and share information. When he was studying Baja's sea turtles as a doctoral student at the University of Arizona in the mid-1990s, he hired fishermen and former turtle poachers to help collect data. The research was interesting, but Nichols was even more intrigued by the intense and often conflicting feelings locals had for the animals. He convened a gathering of everyone —"turtle lovers, turtle eaters, biologists, NGOs"—and they formed an activist network called Grupo Tortuguero.

Nichols was energized, but his academic advisers were skeptical. "'You're organizing fishermen—where's the biology?'" they asked. He was told to avoid the human element in his thesis. "It made no sense," says Nichols. "The changes happening in the ocean and with those turtles were driven by humans."

He was similarly progressive in his research methods. Early on in Baja, he tagged a female turtle his team had named Adelita with a GPS transponder and posted her coordinates online as she made a never-before-recorded crossing of the Pacific to Japan. His colleagues were horrified. "'They said, 'Someone could steal your data!'" Nichols laughs. "My response was 'And do what with it? Save turtles?'"

Today Nichols applies this same open-source spirit to what he calls his "fluid" career. He's spent most of the past decade "hopping between grants" while continuing to publish research on turtles, often coauthored by graduate students he advises. He works with a number of environmental groups and recently created SeetheWild.org, a nonprofit that connects adventure travelers with conservation projects in exotic locations. His office, a 1954 Airstream trailer parked at a friend's organic strawberry farm off California's Highway 1, is also the headquarters for Slowcoast, an initiative he recently helped launch to draw tourists to the mostly empty stretch between Half Moon Bay and Santa Cruz, with future revenue supporting local public school lunch reform.

In 2009 Nichols applied for a grant from the Pew Charitable Trusts to fund a year of neuroscience classes at Harvard and MIT as a way to kick-start his neuroconservation campaign. He posted his twelve-page proposal on his web site the day he submitted it. "I just put it out there," says Nichols. "I was basically saying, Somebody do this, please." Pew turned him down. "They didn't get it," he says. "Which was not a surprise; there's a reason this research hasn't been done."

Indeed, it's one thing to get forward-thinking scientists excited about a hypothesis, but it's another to get institutions to dedicate dollars to test it. Back in 1984, E. O. Wilson popularized the term *biophilia* to describe what he considers humans' inherent attraction to "life and lifelike processes." It became a popular theory but wasn't something Wilson or anyone else initially sought to prove. Now cognitive researchers are investigating what—exactly—nature does to our minds, with studies showing improved attention span and memory and reduced stress, among other benefits. Designing experiments to study how our brains react to the ocean wouldn't be especially difficult, Nichols says. (Among other ideas, he envisions immersing lab subjects in ocean sounds and images while taking brain scans.) But by focusing so explicitly on feelings, Nichols is emanating the kind of New Age vibes that many neuroscientists reflexively avoid. Environmentalists, on the other hand, are prone to question the conservation value of any data such studies might produce. Knowing that something is good for us won't necessarily change our actions (see: exercise, diet, sleep). Plus, what if studies show that a polluted, depleted ocean calms our minds as much as a vibrant one does?

Nichols remains convinced that a researcher will take up his cause soon. Meanwhile, with no regular salary (he isn't paid by the California Academy of Sciences), he has struggled to support his wife, Dana, who manages Slowcoast, and their two grade-school-age daughters while marshaling his neuroconservation drive. His solution is 100BlueAngels.org, a site he established earlier this year that asks people to support him with monthly contributions. Recently he was on pace to bring in what would amount to a $43,000 salary. He supplements this with modeling gigs, which he's taken since college (look for his mug in Gap stores during the holidays), but has had to borrow against his home and take a $10,000 loan from his father.

"People ask me, 'Why don't you sock this idea away until you can get the money and do the research yourself—be the pioneering guy and get all the credit?'" Nichols says. "That's just not as interesting to me. I'd rather hang it out there. Throw a conference. Create the chatter. And hopefully inspire some neuroscientists to ask some of these questions."

Nichols does throw a hell of a conference. This past June, for his Bluemind Summit, which he billed as a gathering that would "forever link the studies of mind and ocean," Nichols wrangled a remarkably eclectic mix of neuro-nerds, greens, adventurers, futurists, artists, a video-game inventor, a high-end realtor, and one very gnarly big-wave surfer to the Academy of Sciences for a marathon day of presentations. The lineup alone demonstrated Nichols's flair for making science both relevant and accessible.

Early on, Eric Johnson, a nattily attired realtor with Sotheby's, cited the premium people are willing to pay for a water view. "We can see the storms or pirates approaching," said Johnson, noting that wealthy owners of high-rise apartments are automatic environmentalists because "clean, clear water keeps property values up." Marcus Eriksen, a marine scientist known for a 2008 crossing of the Pacific in his *Junk,* a raft made primarily of plastic debris, discussed our basic biological reasons for living on the seashore: lots of food and few predators. The ocean activist Fabien Cousteau noted that humans and whales share the mammalian reflex, which allows us to stay underwater for long periods without breathing, while the Maverick's surfer Jeff Clark talked about his learned ability to sense things like the presence of sharks. "Listening to the feedback that the ocean provides will keep you surfing for years," he concluded.

There were some lighter touches. A cellist kicked things off with a medley "full of ocean-ness"—a Nichols request—and each presenter was introduced with a six-word bio ("passion, teacher, vegetables . . ."). At one point, Jaimal Yogis, author of *Saltwater Buddha,* about his quest to find Zen through surfing, led everyone in meditation. Hugs happened.

Still, several cognitive scientists were also on hand to offer serious theories about the brain-on-ocean dynamic. Philippe Goldin, a clinical psychologist and a neuroscientist at Stanford, cited research showing that meditation helped some people with anxiety regain their calm after an emotional event, then speculated that similar processes might be going on in the brains of surfers, who learn to react immediately to a rising swell, then "enjoy the time between waves" after a set passes. Michael Merzenich, a professor emeritus of neuroscience at the University of California at San Francisco and one of the foremost authorities on neuroplasticity—the brain's ability to rewire itself—suggested that our attraction to the ocean may derive from its lack of physical markers. On land we are constantly mapping our environment in our minds so we can pick out dangers (snake!) amid landmarks (tree, bush, rock). Looking over a calm sea is akin to closing our eyes. And when something does emerge on the surface, it captivates us.

Come nightfall, the summit turned into a very northern California kind of happening. We ambled into the academy's planetarium for some ocean-themed readings and a visualization exercise before combining forces with the academy's regular weekly party, which turns the place into a sort of geek-cool dance club. The night's topic was sustainable seafood (purposefully synced with Bluemind), so along with the DJs and organic cocktails there were free local oysters, interactive displays about overfishing and exotic marine creatures, and pamphlets on smarter sushi eating.

Toward the end of the night, on a stage set in a cavernous hall of African wildlife dioramas, a dance troupe in shimmery blue tights and tank tops performed a number called "Aqua." At one point, the oceanographer Sylvia Earle's voice was looped in over house beats: *Imagine an ocean without fish. The ocean is alive.*

Let's say neuroscience does demonstrate that sitting by the ocean provides a unique and primal kind of stimulus that washes away stress. Would our reaction necessarily be a strong desire to protect it? Or might we all instead just selfishly want our own Malibu beach pad?

That sums up the attitude of Michael Soulé, who pioneered the field of conservation biology in the 1970s and later chaired the environmental studies program at the University of California at Santa Cruz. Soulé, as it happens, has been reading neuroscience studies for a book he's writing about our inability to preserve nature. "I admire Nichols's work and his excitement, but the field of cognitive neuroscience can lead you to the opposite conclusions," he told me. Like Nichols, Soulé believes that emotions drive our behavior, but his analysis of fMRI studies, which allow neuroscientists to observe the brain at work, has him convinced that humans are "hard-wired to be very self-centered and self-biased." Understanding why chilling out by the ocean makes us feel great won't motivate a shift in our fundamentally greedy behavior. We are born to be "good consumers but not good conservationists."

Nichols's response: *Of course* we're self-centered. That's why knowing the mechanisms behind something that makes us happy is so powerful—it resonates with our innate desire to feel good, whether we get that feeling from sitting on a beach or protecting it. Regardless, he argues, understanding what goes on in our brains when we're in the presence of the ocean can only help us craft a more persuasive conservation agenda.

Both Soulé and Nichols cite Antonio R. Damasio's popular 1994 book *Descartes' Error* as an influence. Damasio, a neurobiologist, argued that humans can't reason or make decisions without emotion, an idea that ran counter to accepted theories about the division between our rational and emotional selves. Subsequent research showing that the neural pathways of high-level cognition route through our limbic system, the brain's primitive hub where emotions and memories are processed, essentially proved him right. This is why, as marketers and politicians well know, you're most likely to garner dollars or votes by pitching to people's hearts instead of their heads.

The same idea holds true for the environment. At Bluemind, Dawn Martin, the president of SeaWeb, a nonprofit dedicated to strategic messaging on ocean issues, made the case that facts are meaningless unless they're communicated in a way that strikes an emotional chord. She pointed to the collapse of the North Atlantic swordfish population in the mid-1990s, which activists fought unsuccessfully with statistics. Then SeaWeb partnered with the Natural Resources Defense Council to recast the crisis, creating the Give Swordfish a Break campaign, which included an ad with a tiny swordfish on a plate, a pacifier in its mouth: *we're eating the babies.* By 2000 the federal government had closed swordfish nursery areas to fishing.

Nichols appreciates the value of neuroscience-informed social marketing, but he insists that "that's the least interesting part" of what he hopes to learn. "People say, 'Oh, you're just going to be an environmental propagandist.' No. I have no interest in that," he told me. "I want *you* to understand what's happening in your head. My interest is in taking you along for the ride."

Not long ago, I took my two-year-old son to the Academy of Sciences. Almost from the moment we got inside, he was running —past alligators in a pond and snakes in terrariums, up the path that winds around the four-story-tall rain-forest exhibit and its free-flying birds and butterflies, around two life-size model giraffes, all while eating crackers.

Only when we descended the stairs to the Philippine reef did his pace slow. He walked past a couple of smaller windows and came to a halt in front of the massive panoramic view, the cool blue light in the room fluttering about him as shafts of sunlight pierced the water. On one of the benches facing the window, a young mother nursed her infant. People milled about, whispering. My son put his hands on the window and stared without moving or talking for a full thirty seconds—an eternity in toddler time.

Then he spun around, looked me in the eye, and said, "I want to go in there!"

PART FOUR

Humans (the Bad)

THOMAS GOETZ

The Feedback Loop

FROM *Wired*

IN 2003, OFFICIALS in Garden Grove, California, a community of 170,000 people wedged amid the suburban sprawl of Orange County, set out to confront a problem that afflicts almost every town in America: drivers speeding through school zones.

Local authorities had tried many tactics to get people to slow down. They replaced old speed limit signs with bright new ones to remind drivers of the twenty-five-mile-an-hour limit during school hours. Police began ticketing speeding motorists during drop-off and pickup times. But these efforts had only limited success, and speeding cars continued to hit bicyclists and pedestrians in the school zones with depressing regularity.

So city engineers decided to take another approach. In five Garden Grove school zones, they put up what are known as dynamic speed displays, or driver feedback signs: a speed limit posting coupled with a radar sensor attached to a huge digital readout announcing YOUR SPEED.

The signs were curious in a few ways. For one thing, they didn't tell drivers anything they didn't already know—there is, after all, a speedometer in every car. If a motorist wanted to know her speed, a glance at the dashboard would do it. For another thing, the signs used radar, which decades earlier had appeared on American roads as a talisman technology, reserved for police officers only. Now Garden Grove had scattered radar sensors along the side of the road like traffic cones. And the YOUR SPEED signs came with no punitive follow-up—no police officer standing by ready to write a ticket. This defied decades of law enforcement dogma, which held that most people obey speed limits only if they face some clear negative consequence for exceeding them.

In other words, officials in Garden Grove were betting that giving speeders redundant information with no consequence would somehow compel them to do something few of us are inclined to do: slow down.

The results fascinated and delighted the city officials. In the vicinity of the schools where the dynamic displays were installed, drivers slowed an average of 14 percent. Not only that, at three schools the average speed dipped *below* the posted speed limit. Since this experiment, Garden Grove has installed ten more driver feedback signs. "Frankly, it's hard to get people to slow down," says Dan Candelaria, Garden Grove's traffic engineer. "But these encourage people to do the right thing."

In the years since the Garden Grove project began, radar technology has dropped steadily in price, and YOUR SPEED signs have proliferated on American roadways. Yet despite their ubiquity, the signs haven't faded into the landscape like so many other motorist warnings. Instead, they've proven to be consistently effective at getting drivers to slow down—reducing speeds, on average, by about 10 percent, an effect that lasts for several miles down the road. Indeed, traffic engineers and safety experts consider them to be more effective at changing driving habits than a cop with a radar gun. Despite their redundancy, despite their lack of repercussions, the signs have accomplished what seemed impossible: they get us to let up on the gas.

The signs leverage what's called a feedback loop, a profoundly effective tool for changing behavior. The basic premise is simple. Provide people with information about their actions in real time (or something close to it), then give them an opportunity to change those actions, pushing them toward better behaviors. Action, information, reaction. It's the operating principle behind a home thermostat, which fires the furnace to maintain a specific temperature, or the consumption display in a Toyota Prius, which tends to turn drivers into so-called hypermilers, trying to wring every last mile from the gas tank. But the simplicity of feedback loops is deceptive. They are in fact powerful tools that can help people change bad behavior patterns, even those that seem intractable. Just as important, they can be used to encourage good habits, turning progress itself into a reward. In other words, feedback loops change human behavior. And thanks to an explosion of new technology, the opportunity to put them into action in nearly every part of our lives is quickly becoming a reality.

A feedback loop involves four distinct stages. First comes the data: a behavior must be measured, captured, and stored. This is the evidence stage. Second, the information must be relayed to the individual, not in the raw-data form in which it was captured but in a context that makes it emotionally resonant. This is the relevance stage. But even compelling information is useless if we don't know what to make of it, so we need a third stage: consequence. The information must illuminate one or more paths ahead. And finally, the fourth stage: action. There must be a clear moment when the individual can recalibrate a behavior, make a choice, and act. Then that action is measured, and the feedback loop can run once more, every action stimulating new behaviors that inch us closer to our goals.

This basic framework has been shaped and refined by thinkers and researchers for ages. In the eighteenth century, engineers developed regulators and governors to modulate steam engines and other mechanical systems, an early application of feedback loops that later became codified into control theory, the engineering discipline behind everything from aerospace to robotics. The mathematician Norbert Wiener expanded on this work in the 1940s, devising the field of cybernetics, which analyzed how feedback loops operate in machinery and electronics and explored how those principles might be broadened to human systems.

The potential of the feedback loop to affect behavior was explored in the 1960s, most notably in the work of Albert Bandura, a Stanford University psychologist and pioneer in the study of behavior change and motivation. Drawing on several education experiments involving children, Bandura observed that giving individuals a clear goal and a means to evaluate their progress toward that goal greatly increased the likelihood that they would achieve it. He later expanded this notion into the concept of self-efficacy, which holds that the more we believe we can meet a goal, the more likely we will be to do so. In the forty years since Bandura's early work, feedback loops have been thoroughly researched and validated in psychology, epidemiology, military strategy, environmental studies, engineering, and economics. (In typical academic fashion, each discipline tends to reinvent the methodology and rephrase the terminology, but the basic framework remains the same.) Feedback loops are a common tool in athletic training plans, executive coaching strategies, and a multitude of other self-improvement programs (though some are more true to the science than others).

Despite the volume of research and a proven capacity to affect human behavior, we don't often use feedback loops in everyday life. Blame this on two factors. Until now, the necessary catalyst —personalized data—has been an expensive commodity. Health spas, athletic training centers, and self-improvement workshops all traffic in fastidiously culled data at premium rates. Outside of those rare realms, the cornerstone information has been just too expensive to come by. As a technologist might put it, personalized data hasn't really scaled.

Second, collecting data on the cheap is cumbersome. Although the basic idea of self-tracking has been available to anyone willing to put in the effort, few people stick with the routine of toting around a notebook, writing down every Hostess cupcake they consume or every flight of stairs they climb. It's just too much bother. The technologist would say that capturing that data involves too much friction. As a result, feedback loops are niche tools for the most part, rewarding for those with the money, willpower, or geeky inclination to obsessively track their own behavior, but impractical for the rest of us.

That's quickly changing because of one essential technology: sensors. Adding sensors to the feedback equation helps solve problems of friction and scale. They automate the capture of behavioral data, digitizing it so that it can be readily crunched and transformed as necessary. And they allow passive measurement, eliminating the need for tedious active monitoring.

In the past two or three years, the plunging price of sensors has begun to foster a feedback-loop revolution. Just as YOUR SPEED signs have been adopted worldwide because the cost of radar technology keeps dropping, other feedback loops are popping up everywhere because sensors keep getting cheaper and better at monitoring behavior and capturing data in all sorts of environments. These new, less expensive devices include accelerometers (which measure motion), GPS sensors (which track location), and inductance sensors (which measure electric current). Accelerometers have dropped to less than a dollar each—down from as much as twenty dollars a decade ago—which means they can now be built into tennis shoes, MP3 players, and even toothbrushes. Radio-frequency ID chips are being added to prescription pill bottles, student ID cards, and casino chips. And inductance sensors that were once deployed only in heavy industry are now cheap and tiny enough to be connected to residential breaker boxes, letting consumers track their home's entire energy diet.

Of course, technology has been tracking what people do for years. Call-center agents have been monitored closely since the 1990s, and the nation's tractor-trailer fleets have long been equipped with GPS and other location sensors—not just to allow drivers to follow their routes but to enable companies to track their cargo and the drivers. But those are top-down, Big Brother techniques. The true power of feedback loops is not to control people but to give them control. It's like the difference between a speed trap and a speed feedback sign—one is a game of gotcha, the other is a gentle reminder of the rules of the road. The ideal feedback loop gives us an emotional connection to a rational goal.

And today their promise couldn't be greater. The intransigence of human behavior has emerged as the root of most of the world's biggest challenges. Witness the rise in obesity, the persistence of smoking, the soaring number of people who have one or more chronic diseases. Consider our problems with carbon emissions, where managing personal energy consumption could be the difference between a climate under control and one beyond help. And feedback loops aren't just about solving problems. They could create opportunities. Feedback loops can improve how companies motivate and empower their employees, allowing workers to monitor their own productivity and set their own schedules. They could lead to lower consumption of precious resources and more productive use of what we do consume. They could allow people to set and achieve better-defined, more ambitious goals and curb destructive behaviors, replacing them with positive actions. Used in organizations or communities, they can help groups work together to take on more daunting challenges. In short, the feedback loop is an age-old strategy revitalized by state-of-the-art technology. As such, it is perhaps the most promising tool for behavioral change to have come along in decades.

*

In 2006 Shwetak Patel, then a graduate student in computer science at Georgia Tech, was working on a problem: How could technology help provide remote care for the elderly? The obvious approach would be to install cameras and motion detectors throughout a home, so that observers could see when somebody fell or became sick. Patel found those methods unsophisticated and impractical. "Installing cameras or motion sensors everywhere is unreasonably expensive," he says. "It might work in theory, but it just won't happen in practice. So I wondered what would give us the same information and be reasonably priced and easy to deploy. I found those really interesting constraints."

The answer, Patel realized, is that every home emits something called voltage noise. Think of it as a steady hum in the electrical wires that varies depending on what systems are drawing power. If there was some way to disaggregate this noise, it might be possible to deliver much the same information as cameras and motion sensors can. Lights going on and off, for instance, would mean that someone had moved from room to room. If a blender was left on, that might signal that someone had fallen—or had forgotten about the blender, perhaps indicating dementia. If we could hear electricity usage, Patel thought, we could know what was happening inside the house.

A nifty idea, but how to make it happen? The problem wasn't measuring the voltage noise; that's easily tracked with a few sensors. The challenge was translating the cacophony of electromagnetic interference into the symphony of signals given off by specific appliances and devices and lights. Finding that pattern amid the noise became the focus of Patel's PhD work, and in a few years he had both his degree and his answer: a stack of algorithms that could discern a blender from a light switch from a television set and so on. All this data could be captured not by sensors in every electrical outlet throughout the house but through a single device plugged into a single outlet.

This, Patel soon realized, went way beyond elder care. His approach could inform ordinary consumers, in real time, about where the energy they paid for every month was going. "We kind of stumbled across this stuff," Patel says. "But we realized that combined with data on the house's overall draw on power"—which can be measured through a second sensor easily installed at the circuit box—"we were getting really great information about resource consumption in the home. And that could be more than interesting information. It could encourage behavior change."

By 2008 Patel had started a new job in the computer science and engineering departments at the University of Washington, and his idea had been turned into the startup Zensi. At the university he focused on devising similar techniques to monitor home consumption of water and gas. The solutions were even more elegant, perhaps, than the one for monitoring electricity. A transducer affixed to an outdoor spigot can detect changes in water pressure that correspond to the resident's water usage. That data can then be disaggregated to distinguish a leaky toilet from an overindulgent bather. And a microphone sensor on a gas meter listens to changes in the regulator to determine how much gas is consumed.

Last year the consumer-electronics company Belkin acquired Zensi and made energy conservation a centerpiece of its corporate strategy, with feedback loops as the guiding principle. Belkin has begun modestly with a device called the Conserve Insight. It's an outlet adapter that gives consumers a close read of the power used by one select appliance: plug it into a wall socket and then plug an appliance or gadget into it, and a small display shows how much energy the device is consuming, in both watts and dollars. It's a window onto how energy is actually used, but it's only a proof-of-concept prototype of the more ambitious product, based on Patel's PhD work, which Belkin will begin beta-testing in Chicago later this year with an eye toward commercial release in 2013. The company calls it Zorro.

At first glance, the Zorro is just another so-called smart meter, not that different from the boxes that many power companies have been installing in consumers' homes, with a vague promise that the meters will educate citizens and provide better data to the utility. To the surprise of the utility companies, though, these smart meters have been greeted with hostility in some communities. A small but vocal number of customers object to being monitored, while others worry that the radiation from RFID (radio-frequency identification) transmitters is unhealthy (though this has been measured at infinitesimal levels).

Politics aside, in pure feedback terms smart meters fail on at least two levels. For one, the information goes to the utility first rather than directly to the consumer. For another, most smart meters aren't very smart; they typically measure overall household consumption, not how much power is being consumed by which specific device or appliance. In other words, they are a broken feedback loop.

Belkin's device avoids these pitfalls by giving the data directly to consumers and delivering it promptly and continuously. "Real-time feedback is key to conservation," says Kevin Ashton, Zensi's former CEO, who took over Belkin's Conserve division after the acquisition. "There's a visceral impact when you see for yourself how much your toaster is costing you."

The Zorro is just the first of several Belkin products that Ashton believes will put feedback loops into effect throughout the home. Ashton worked on RFID chips at MIT in the late 1990s and lays claim to coining the phrase "Internet of Things," meaning a world of interconnected, sensor-laden devices and objects. He predicts that home sensors will one day inform choices in all aspects of our lives. "We're consuming so many things without thinking about them—energy, plastic, paper, calories. I can envision a ubiquitous sensor network, a platform for real-time feedback that will enhance the comfort, security, and control of our lives."

As a starting point for a consumer-products company, that's not half bad.

If there is one problem in medicine that confounds doctors, insurers, and pharmaceutical companies alike, it's noncompliance, the unfriendly term for patients' not following doctors' orders. Most vexing are patients who don't take their medications as prescribed—which, it turns out, is pretty much most of us. Studies have shown that about half of patients who are prescribed medication take their pills as directed. For drugs like statins, which must be used for years, the rate is even worse, dropping to around 30 percent after a year. (Since the effects of these drugs can be invisible, the thinking goes, patients don't detect any benefit from them.) Research has found that noncompliance adds $100 billion annually to US health care costs and leads to 125,000 unnecessary deaths from cardiovascular diseases alone every year. And it can be blamed almost entirely on human foibles—people failing to do what they know they should.

David Rose is a perfect example of this. He has a family history of heart disease. Now forty-four, he began taking medication for high blood pressure a few years ago, making him not so different from the nearly one-third of Americans with hypertension. Where Rose is exceptional is in his capacity to do something about noncompliance. He has a knack for inventing beautiful, engaging, alluring objects that get people to do things like take their pills.

A decade ago, Rose, whose stylish glasses and soft-spoken manner bring to mind a college music teacher, started a company called Ambient Devices. His most famous product is the Orb, a translucent sphere that turns different colors to reflect different information inputs. If your stocks go down, it might glow red; if it snows, it might glow white, and so on, depending on what information you tell the Orb you are interested in. It's a whimsical product and is still available for purchase online. But as far as Rose is concerned, the Orb was merely a prelude to his next company, Vitality, and its marquee product: the GlowCap.

The device is simple. When a patient is prescribed a medication, a physician or pharmacy provides a GlowCap to go on top of the pill bottle, replacing the standard childproof cap. The GlowCap, which comes with a plug-in unit that Rose calls a night-light, connects to a database that knows the patient's particular dosage directions—say, two pills twice a day, at eight A.M. and eight P.M. When eight A.M. rolls around, the GlowCap and the night-light start to pulse with a gentle orange light. A few minutes later, if the pill bottle isn't opened, the light pulses a little more urgently. A few minutes more and the device begins to play a melody—not an annoying buzz or alarm. Finally, if more time elapses (the intervals are adjustable), the patient receives a text message or a recorded phone call reminding him to pop the GlowCap. The overall effect is a persistent feedback loop urging patients to take their meds.

These nudges have proven to be remarkably effective. In 2010 Partners HealthCare and Harvard Medical School conducted a study that gave GlowCaps to 140 patients on hypertension medications; a control group received nonactivated GlowCap bottles. After three months, adherence in the control group had declined to less than 50 percent, the same dismal rate observed in countless other studies. But patients using GlowCaps did remarkably better: more than 80 percent of them took their pills, a rate that lasted for the duration of the six-month study.

The power of the device can perhaps be explained by the

fact that the GlowCap incorporates several schools of behavioral change. Vitality has experimented with charging consumers for the product, drawing on the behavioral-economics theory that people are more willing to use something they've paid for. But in other circumstances the company has given users a financial reward for taking their medication, using a carrot-and-stick methodology. Different models work for different people, Rose says. "We use reminders and social incentives and financial incentives —whatever we can," he says. "We want to provide enough feedback so that it's complementary to people's lives, but not so much that you can't handle the onslaught."

Here Rose grapples with an essential challenge of feedback loops. Make them too passive and you'll lose your audience as the data blurs into the background of everyday life. Make them too intrusive and the data turns into noise, which is easily ignored. Borrowing a concept from cognitive psychology called pre-attentive processing, Rose aims for a sweet spot between these extremes, where the information is delivered unobtrusively but noticeably. The best sort of delivery device "isn't cognitively loading at all," he says. "It uses colors, patterns, angles, speed—visual cues that don't distract us but remind us." This creates what Rose calls "enchantment." Enchanted objects, he says, don't register as gadgets or even as technology at all, but rather as friendly tools that beguile us into action. In short, they're magical.

This approach to information delivery is a radical departure from how our health care system usually works. Conventional wisdom holds that medical information won't be heeded unless it sets off alarms. Instead of glowing orbs, we're pummeled with FDA cautions and Surgeon General warnings and front-page reports, all of which serve to heighten our anxiety about our health. This fear-based approach can work—for a while. But fear, it turns out, is a poor catalyst for sustained behavioral change. After all, biologically our fear response girds us for short-term threats. If nothing threatening actually happens, the fear dissipates. If this happens too many times, we end up simply dismissing the alarms.

It's worth noting here how profoundly difficult it is for most people to improve their health. Consider: self-directed smoking-cessation programs typically work for perhaps 5 percent of participants, and weight-loss programs are considered effective if people lose as little as 5 percent of their body weight. Part of the problem is that so much in our lives—the foods we eat, the ads we see, the things our culture celebrates—is driven by negative feedback loops that sustain bad behaviors. A positive feedback loop offers a chance to counterprogram this onslaught and dramatically increase our odds of changing course.

Though GlowCaps improved compliance by an astonishing 40 percent, feedback loops more typically improve outcomes by about 10 percent compared to traditional methods. That 10 percent figure is surprisingly persistent; it turns up in everything from home energy monitors to smoking cessation programs to those YOUR SPEED signs. At first glance, 10 percent may not seem like a lot. After all, if you're 250 pounds and obese, losing 25 pounds is a start, but your BMI (body mass index) is likely still in the red zone. But it turns out that 10 percent does matter. A lot. An obese forty-year-old man would spare himself three years of hypertension and nearly two years of diabetes by losing 10 percent of his weight. A 10 percent reduction in home energy consumption could reduce carbon emissions by as much as 20 percent (generating energy during peak demand periods creates more pollution than off-peak generation). And those YOUR SPEED signs? It turns out that reducing speeds by 10 percent, from forty to thirty-five miles an hour, would cut fatal injuries by about half.

In other words, 10 percent is something of an inflection point, where lots of great things happen. The results are measurable, the economics calculable. "The value of behavior change is incredibly large: nearly five thousand dollars a year," says David Rose, citing a CVS pharmacy white paper. "At that rate, we can afford to give every diabetic a connected glucometer. We can give the morbidly obese a Wi-Fi–enabled scale and a pedometer. The value is there; the savings are there. The cost of the sensors is negligible."

So feedback loops work. Why? Why does putting our own data in front of us somehow compel us to act? In part, it's that feedback taps into something core to the human experience, even to our biological origins. Like any organism, humans are self-regulating creatures, with a multitude of systems working to achieve homeostasis. Evolution itself, after all, is a feedback loop, albeit one so elongated as to be imperceptible by an individual. Feedback loops are how we learn, whether we call it trial and error or course correction. In so many areas of life, we succeed when we have some sense of where we stand and some evaluation of our progress. Indeed, we tend to crave this sort of information; it's something we viscerally want to know, good or bad. As Stanford's Bandura put it, "People are proactive, aspiring organisms." Feedback taps into those aspirations.

The visceral satisfaction and even pleasure we get from feedback loops is the organizing principle behind GreenGoose, a startup being hatched by Brian Krejcarek, a Minnesota native who wears a near-constant smile, so enthusiastic is he about the power of cheap sensors. His mission is to stitch feedback loops into the fabric of our daily lives, one sensor at a time.

As Krejcarek describes it, GreenGoose started with a goal not too different from Shwetak Patel's: to measure household consumption of energy. But the company's mission took a turn in 2009, when he experimented with putting one of those ever-cheaper accelerometers on a bicycle wheel. As the wheel rotated, the sensor picked up the movement, and before long Krejcarek had a vision of a grander plan. "I wondered what else we could measure. Where else could we stick these things?" The answer he came up with: everywhere. The GreenGoose concept starts with a sheet of stickers, each containing an accelerometer labeled with a cartoon icon of a familiar household object—a refrigerator handle, a water bottle, a toothbrush, a yard rake. But the secret to GreenGoose isn't the accelerometer; that's a less-than-a-dollar commodity. The key is the algorithm that Krejcarek's team has coded into the chip next to the accelerometer that recognizes a particular pattern of movement. For a toothbrush, it's a rapid back-and-forth that indicates somebody is brushing her teeth. For a water bottle, it's a simple up-and-down that correlates with somebody taking a sip. And so on. In essence, GreenGoose uses sensors to spray feedback loops like atomized perfume throughout our daily life—in our homes, our vehicles, our backyards. "Sensors are these little eyes and ears on whatever we do and how we do it," Krejcarek says. "If a behavior has a pattern, if we can calculate a desired duration and intensity, we can create a system that rewards that behavior and encourages more of it." Thus the first component of a feedback loop: data gathering.

Then comes the second step: relevance. GreenGoose converts the data into points, with a certain amount of action translating into a certain number of points, say, thirty seconds of teeth brushing for two points. And here Krejcarek gets noticeably excited. "The points can be used in games on our web site," he says. "Think FarmVille but with live data." Krejcarek plans to open the platform to game developers, who he hopes will create games that are simple, easy, and sticky. A few hours of raking leaves might build up points that can be used in a gardening game. And the games induce people to earn more points, which means repeating good behaviors. The idea, Krejcarek says, is to "create a bridge between the real world and the virtual world. This has all got to be fun."

As powerful as the idea appears now, not long ago it seemed like a fading pipe dream. Then based in Cambridge, Massachusetts, Krejcarek had nearly run out of cash—not just for his company but for himself. During the day, he was working on GreenGoose in an office building near the MIT campus—and each night, he'd sneak into the building's air shaft, where he'd stashed an air mattress and some clothes. Then, in late February 2011, he went to the Launch conference in San Francisco, a two-day event where select entrepreneurs get a chance to demo their company to potential funders. Krejcarek hadn't been selected for an onstage demo, but when the conference organizers saw a crowd eyeing his product on the exhibit floor, he was given four minutes to make a presentation. It was one of those only-in-Silicon Valley moments. The crowd "just got it," he recalls. Within days, he had nearly $600,000 in new funding. He moved to San Francisco, rented an apartment—and bought a bed. GreenGoose will release its first product, a kit of sensors that encourage pet owners to play and interact with their dogs, with sensors for dog collars, pet toys, and dog doors, sometime this fall.

Part of the excitement around GreenGoose is that the company is so good at "gamification," the much-blogged-about notion that game elements like points or levels can be applied to various aspects of our lives. Gamification is exciting because it promises to make the hard stuff in life fun—just sprinkle a little video-game magic and suddenly a burden turns into bliss. But as happens with fads, gamification is both overhyped and misunderstood. It is too often just a shorthand for badges or points, like so many gold stars on a spelling test. But just as no number of gold stars can trick children into thinking that yesterday's quiz was fun, game mechanics, to work, must be an informing principle, not a veneer.

With its savvy application of feedback loops, though, Green-

Goose is on to more than just the latest fad. The company represents the fruition of a long-promised technological event horizon: the Internet of Things, in which a sensor-rich world measures our every action. This vision, championed by Kevin Ashton at Belkin, Sandy Pentland at MIT, and Bruce Sterling in the pages of *Wired*, has long had the whiff of vaporware, something promised by futurists but never realized. But as GreenGoose, Belkin, and other companies begin to use sensors to deploy feedback loops throughout our lives, we can finally see the potential of a sensor-rich environment. The Internet of Things isn't about the things; it's about us.

For now, the reality still isn't as sexy as the visions. Stickers on toothbrushes and plugs in wall sockets aren't exactly disappearing technology. But maybe requiring people to do a little work—to stick accelerometers around their house or plug a device into a wall socket—is just enough of a nudge to get our brains engaged in the prospect for change. Perhaps it's good to have the infrastructure of feedback loops just a bit visible now, before they disappear into our environments altogether, so that they can serve as a subtle reminder that we have something to change, that we can do better—and that the tools for doing better are rapidly, finally, turning up all around us.

JASON DALEY

What You Don't Know Can Kill You

FROM *Discover*

IN MARCH 2011, as the world watched the aftermath of the Japanese earthquake/tsunami/nuclear near-meltdown, a curious thing began happening in West Coast pharmacies. Bottles of potassium iodide pills used to treat certain thyroid conditions were flying off the shelves, creating a run on an otherwise obscure nutritional supplement. Online prices jumped from ten dollars a bottle to upward of two hundred dollars. Some residents in California, unable to get the iodide pills, began bingeing on seaweed, which is known to have high iodine levels.

The Fukushima disaster was practically an infomercial for iodide therapy. The chemical is administered after nuclear exposure because it helps protect the thyroid from radioactive iodine, one of the most dangerous elements of nuclear fallout. Typically, iodide treatment is recommended for residents within a ten-mile radius of a radiation leak. But the people in the United States who were popping pills were at least five thousand miles away from the Japanese reactors. Experts at the Environmental Protection Agency estimated that the dose of radiation that reached the western United States was equivalent to 1/100,000 the exposure one would get from a round-trip international flight.

Although spending two hundred dollars on iodide pills for an almost nonexistent threat seems ridiculous (and could even be harmful—side effects include skin rashes, nausea, and possible allergic reactions), forty years of research into the way people perceive risk shows that it is par for the course. Earthquakes? Tsunamis? Those things seem inevitable, accepted as acts of God. But an invisible, man-made threat associated with Godzilla and three-eyed fish? Now that's something to keep you up at night. "There's a lot of emotion that comes from the radiation in Japan," says the cognitive psychologist Paul Slovic, an expert on decision making and risk assessment at the University of Oregon. "Even though the earthquake and tsunami took all the lives, all of our attention was focused on the radiation."

We like to think that humans are supremely logical, making decisions on the basis of hard data and not on whim. For a good part of the nineteenth and twentieth centuries, economists and social scientists assumed this was true too. The public, they believed, would make rational decisions if only it had the right pie chart or statistical table. But in the late 1960s and early 1970s, that vision of *homo economicus*—a person who acts in his or her best interest when given accurate information—was kneecapped by researchers investigating the emerging field of risk perception. What they found, and what they have continued teasing out since the early 1970s, is that humans have a hell of a time accurately gauging risk. Not only do we have a system that gives us conflicting advice from two powerful sources—logic and instinct, or the head and the gut—but we are also at the mercy of deep-seated emotional associations and mental shortcuts.

Even if a risk has an objectively measurable probability—like the chances of dying in a fire, which are 1 in 1,177—people will assess the risk subjectively, mentally calibrating it based on dozens of subconscious calculations. If you have been watching news coverage of wildfires in Texas nonstop, chances are you will assess the risk of dying in a fire higher than will someone who has been floating in a pool all day. If the day is cold and snowy, you are less likely to think global warming is a threat.

Our hard-wired gut reactions developed in a world full of hungry beasts and warring clans, where they served important functions. Letting the amygdala (part of the brain's emotional core) take over at the first sign of danger, milliseconds before the neocortex (the thinking part of the brain) was aware that a spear was headed for our chest, was probably a very useful adaptation. Even today those nano-pauses and gut responses save us from getting flattened by buses or dropping a brick on our toes. But in a world where risks are presented in parts-per-billion statistics or as clicks on a Geiger counter, our amygdala is out of its depth.

A risk-perception apparatus permanently tuned for avoiding mountain lions makes it unlikely that we will ever run screaming from a plate of fatty mac 'n' cheese. "People are likely to react with little fear to certain types of objectively dangerous risk that evolution has not prepared them for, such as guns, hamburgers, automobiles, smoking, and unsafe sex, even when they recognize the threat at a cognitive level," says the Carnegie Mellon University researcher George Loewenstein, whose seminal 2001 paper, "Risk as Feelings," debunked theories that decision making in the face of risk or uncertainty relies largely on reason. "Types of stimuli that people are evolutionarily prepared to fear, such as caged spiders, snakes, or heights, evoke a visceral response even when, at a cognitive level, they are recognized to be harmless," he says. Even Charles Darwin failed to break the amygdala's iron grip on risk perception. As an experiment, he placed his face up against the puff adder enclosure at the London Zoo and tried to keep himself from flinching when the snake struck the plate glass. He failed.

The result is that we focus on the one-in-a-million bogeyman while virtually ignoring the true risks that inhabit our world. News coverage of a shark attack can clear beaches all over the country, even though sharks kill a grand total of about one American annually, on average. That is less than the death count from cattle, which gore or stomp 20 Americans per year. Drowning, on the other hand, takes 3,400 lives a year, without a single frenzied call for mandatory life vests to stop the carnage. A whole industry has boomed around conquering the fear of flying, but while we down beta-blockers in coach, praying not to be one of the 48 average annual airline casualties, we typically give little thought to driving to the grocery store, even though there are more than 30,000 automobile fatalities each year.

In short, our risk perception is often at direct odds with reality. All those people bidding up the cost of iodide? They would have been better off spending ten dollars on a radon testing kit. The colorless, odorless, radioactive gas, which forms as a byproduct of natural uranium decay in rocks, builds up in homes, causing lung cancer. According to the Environmental Protection Agency, radon exposure kills 21,000 Americans annually.

David Ropeik, a consultant in risk communication and the author of *How Risky Is It, Really? Why Our Fears Don't Always Match the Facts,* has dubbed this disconnect the perception gap. "Even perfect information perfectly provided that addresses people's concerns will not convince everyone that vaccines don't cause autism, or that global warming is real, or that fluoride in the drinking water is not a Commie plot," he says. "Risk communication can't totally close the perception gap, the difference between our fears and the facts."

In the early 1970s, the psychologists Daniel Kahneman, now at Princeton University, and Amos Tversky, who passed away in 1996, began investigating the way people make decisions, identifying a number of biases and mental shortcuts, or heuristics, on which the brain relies to make choices. Later Paul Slovic and his colleagues Baruch Fischhoff, now a professor of social sciences at Carnegie Mellon University, and psychologist Sarah Lichtenstein began investigating how these leaps of logic come into play when people face risk. They developed a tool, called the psychometric paradigm, that describes all the little tricks our brain uses when staring down a bear or deciding to finish the eighteenth hole in a lighting storm.

Many of our personal biases are unsurprising. For instance, the optimism bias gives us a rosier view of the future than current facts might suggest. We assume we will be richer ten years from now, so it is fine to blow our savings on a boat—we'll pay it off then. Confirmation bias leads us to prefer information that backs up our current opinions and feelings and to discount information contradictory to those opinions. We also have tendencies to conform our opinions to those of the groups we identify with, to fear man-made risks more than we fear natural ones, and to believe that events causing dread—the technical term for risks that could result in particularly painful or gruesome deaths, like plane crashes and radiation burns—are inherently more risky than other events.

But it is heuristics—the subtle mental strategies that often give rise to such biases—that do much of the heavy lifting in risk perception. The "availability" heuristic says that the easier a scenario is to conjure, the more common it must be. It is easy to imagine a tornado ripping through a house; that is a scene we see every spring on the news and all the time on reality TV and in movies.

Now try imagining someone dying of heart disease. You probably cannot conjure many breaking-news images for that one, and the drawn-out process of atherosclerosis will most likely never be the subject of a summer thriller. The effect? Twisters feel like an immediate threat, although we have only a 1-in-46,000 chance of being killed by a cataclysmic storm. Even a terrible tornado season like the one last spring typically yields fewer than 500 tornado fatalities. Heart disease, on the other hand, which eventually kills 1 in every 4 people in this country, and 800,000 annually, hardly even rates with our gut.

The "representative" heuristic makes us think something is probable if it is part of a known set of characteristics. John wears glasses, is quiet, and carries a calculator. John is therefore . . . a mathematician? An engineer? His attributes taken together seem to fit the common stereotype.

But all of those mental rules of thumb and biases banging around our brain are only part of a larger risk-perception system. The affect heuristic, a type of unified-field theory of risk perception, emcompasses those biases, and also hints at other factors researchers are slowly untangling, which may be even more influential in the way we make choices. Slovic, who recently edited *The Feeling of Risk: New Perspectives on Risk Perception,* calls affect a "faint whisper of emotion" that creeps into our decisions. In fact, just reading words like *radiation* or *head trauma* creates a split second of emotion that can subconsciously influence us. His research and studies by others have shown that positive feelings associated with a choice tend to make us think it has more benefits. Negative correlations make us think an action is riskier. One study by Slovic showed that when people decide to start smoking despite years of exposure to antismoking campaigns, they hardly ever think about the risks. Instead, it's all about the short-term "hedonic" pleasure. The good outweighs the bad, which they never fully expect to experience.

Our fixation on illusory threats at the expense of real ones influences more than just our personal lifestyle choices. Public policy and mass action are also at stake. The Office of National Drug Control Policy reports that prescription drug overdoses have killed more people than crack and heroin combined did in the 1970s and 1980s. Law enforcement and the media were obsessed with crack, yet it was only recently that prescription drug abuse merited even an after-school special.

Despite the many obviously irrational ways we behave, social scientists have only just begun to systematically document and understand this central aspect of our nature. In the 1960s and 1970s, many still clung to the *Homo economicus* model. They argued that releasing detailed information about nuclear power and pesticides would convince the public that these industries were safe. But the information drop was an epic backfire and helped spawn opposition groups that exist to this day. Part of the resistance stemmed from a reasonable mistrust of industry spin. Horrific incidents like those at Love Canal and Three Mile Island did not help. Yet one of the biggest obstacles was that industry tried to frame risk purely in terms of data, without addressing the fear that is an instinctual reaction to their technologies.

The strategy persists even today. In the aftermath of Japan's nuclear crisis, many nuclear-energy boosters were quick to cite a study commissioned by the Boston-based nonprofit Clean Air Task Force. The study showed that pollution from coal plants is responsible for 13,000 premature deaths and 20,000 heart attacks in the United States each year, while nuclear power has never been implicated in a single death in this country. True as that may be, numbers alone cannot explain away the cold dread caused by the specter of radiation. Just think of all those alarming images of workers clad in radiation suits waving Geiger counters over the anxious citizens of Japan. Seaweed, anyone?

At least a few technology promoters have become much more savvy in understanding the way the public perceives risk. The nanotechnology world in particular has taken a keen interest in this process, since even in its infancy it has faced high-profile fears. Nanotech, a field so broad that even its backers have trouble defining it, deals with materials and devices whose components are often smaller than 1/100,000,000,000 of a meter. In the late 1980s, the book *Engines of Creation* by the nanotechnologist K. Eric Drexler put forth the terrifying idea of nanoscale self-replicating robots that grow into clouds of "gray goo" and devour the world. Soon gray goo was turning up in video games, magazine stories, and delightfully bad Hollywood action flicks (see, for instance, the last G.I. Joe movie).

The odds of nanotechnology's killing off humanity are extremely remote, but the science is obviously not without real risks. In 2008 a study led by researchers at the University of Edinburgh suggested that carbon nanotubes, a promising material that could be used in everything from bicycles to electrical circuits, might interact with the body the same way asbestos does. In another study, scientists at the University of Utah found that nanoscopic particles of silver used as an antimicrobial in hundreds of products, including jeans, baby bottles, and washing machines, can deform fish embryos.

The nanotech community is eager to put such risks in perspective. "In Europe, people made decisions about genetically modified food irrespective of the technology," says Andrew Maynard, director of the Risk Science Center at the University of Michigan and an editor of the *International Handbook on Regulating Nanotechnologies*. "People felt they were being bullied into the technology by big corporations, and they didn't like it. There have been very small hints of that in nanotechnology." He points to incidents in which sunblock makers did not inform the public that they were including zinc oxide nanoparticles in their products, stoking the skepticism and fears of some consumers.

For Maynard and his colleagues, influencing public perception has been an uphill battle. A 2007 study conducted by the Cultural Cognition Project at Yale Law School and coauthored by Paul Slovic surveyed 1,850 people about the risks and benefits of nanotech. Even though 81 percent of participants knew nothing or very little about nanotechnology before starting the survey, 89 percent of all respondents said they had an opinion on whether nanotech's benefits outweighed its risks. In other words, people made a risk judgment based on factors that had little to do with any knowledge about the technology itself. And as with public reaction to nuclear power, more information did little to unite opinions. "Because people with different values are predisposed to draw different factual conclusions from the same information, it cannot be assumed that simply supplying accurate information will allow members of the public to reach a consensus on nanotechnology risks, much less a consensus that promotes their common welfare," the study concluded.

It should come as no surprise that nanotech hits many of the fear buttons in the psychometric paradigm: it is a man-made risk;

much of it is difficult to see or imagine; and the only available images we can associate with it are frightening movie scenes, such as a cloud of robots eating the Eiffel Tower. "In many ways, this has been a grand experiment in how to introduce a product to the market in a new way," Maynard says. "Whether all the up-front effort has gotten us to a place where we can have a better conversation remains to be seen."

That job will be immeasurably more difficult if the media—in particular cable news—ever decide to make nanotech their fear du jour. In the summer of 2001, if you switched on the television or picked up a newsmagazine, you might think the ocean's top predators had banded together to take on humanity. After eight-year-old Jessie Arbogast's arm was severed by a seven-foot bull shark on the Fourth of July weekend while the child was playing in the surf off Santa Rosa Island, near Pensacola, Florida, cable news put all its muscle behind the story. Ten days later, a surfer was bitten just six miles from the beach where Jessie had been mauled. Then a lifeguard in New York claimed he had been attacked. There was almost round-the-clock coverage of the "Summer of the Shark," as it came to be known. By August, according to an analysis by the historian April Eisman of Iowa State University, it was the third-most-covered story of the summer until the September 11 attacks knocked sharks off the cable news channels.

All that media created a sort of feedback loop. Because people were seeing so many sharks on television and reading about them, the "availability" heuristic was screaming at them that sharks were an imminent threat.

"Certainly anytime we have a situation like that where there's such overwhelming media attention, it's going to leave a memory in the population," says George Burgess, curator of the International Shark Attack File at the Florida Museum of Natural History, who fielded thirty to forty media calls a day that summer. "Perception problems have always been there with sharks, and there's a continued media interest in vilifying them. It makes a situation where the risk perceptions of the populace have to be continually worked on to break down stereotypes. Anytime there's a big shark event, you take a couple steps backward, which requires scientists and conservationists to get the real word out."

*

Then again, getting out the real word comes with its own risks —like the risk of getting the real word wrong. Misinformation is especially toxic to risk perception because it can reinforce generalized confirmation biases and erode public trust in scientific data. As scientists studying the societal impact of the Chernobyl meltdown have learned, doubt is difficult to undo. In 2006, twenty years after reactor number 4 at the Chernobyl nuclear power plant was encased in cement, the World Health Organization (WHO) and the International Atomic Energy Agency released a report compiled by a panel of one hundred scientists on the long-term health effects of the level-7 nuclear disaster and future risks for those exposed. Among the 600,000 recovery workers and local residents who received a significant dose of radiation, the WHO estimates that up to 4,000 of them, or 0.7 percent, will develop a fatal cancer related to Chernobyl. For the 5 million people living in less contaminated areas of Ukraine, Russia, and Belarus, radiation from the meltdown is expected to increase cancer rates less than 1 percent.

Even though the percentages are low, the numbers are little comfort for the people living in the shadow of the reactor's cement sarcophagus who are literally worrying themselves sick. In the same report, the WHO states that "the mental health impact of Chernobyl is the largest problem unleashed by the accident to date," pointing out that fear of contamination and uncertainty about the future have led to widespread anxiety, depression, hypochondria, alcoholism, a sense of victimhood, and a fatalistic outlook that is extreme even by Russian standards. A recent study in the journal *Radiology* concludes that "the Chernobyl accident showed that overestimating radiation risks could be more detrimental than underestimating them. Misinformation partially led to traumatic evacuations of about 200,000 individuals, an estimated 1,250 suicides, and between 100,000 and 200,000 elective abortions."

It is hard to fault the Chernobyl survivors for worrying, especially when it took twenty years for the scientific community to get a grip on the aftereffects of the disaster, and even those numbers are disputed. An analysis commissioned by Greenpeace in response to the WHO report predicts that the Chernobyl disaster will result in about 270,000 cancers and 93,000 fatal cases.

Chernobyl is far from the only chilling illustration of what can happen when we get risk wrong. During the year following the September 11 attacks, millions of Americans opted out of air travel and slipped behind the wheel instead. While they crisscrossed the country, listening to breathless news coverage of anthrax attacks, extremists, and Homeland Security, they faced a much more concrete risk. All those extra cars on the road increased traffic fatalities by nearly 1,600. Airlines, on the other hand, recorded no fatalities.

It is unlikely that our intellect can ever paper over our gut reactions to risk. But a fuller understanding of the science is beginning to percolate into society. Earlier this year, David Ropeik and others hosted a conference on risk in Washington, DC, bringing together scientists, policymakers, and others to discuss how risk perception and communication impact society. "Risk perception is not emotion and reason, or facts and feelings. It's both, inescapably, down at the very wiring of our brain," says Ropeik. "We can't undo this. What I heard at that meeting was people beginning to accept this and to realize that society needs to think more holistically about what risk means."

Ropeik says policymakers need to stop issuing reams of statistics and start making policies that manage our risk-perception system instead of trying to reason with it. Cass Sunstein, a Harvard law professor who is now the administrator of the White House Office of Information and Regulatory Affairs, suggests a few ways to do this in his book *Nudge: Improving Decisions about Health, Wealth, and Happiness,* published in 2008. He points to the organ donor crisis, in which thousands of people die each year because others are too fearful or uncertain to donate organs. People tend to believe that doctors won't work as hard to save them or that they won't be able to have an open-casket funeral (both false). And the gory mental images of organs being harvested from a body give a definite negative affect to the exchange. As a result, too few people focus on the lives that could be saved. Sunstein suggests—controversially—"mandated choice," in which people must check "yes" or "no" to organ donation on their driver's license application. Those with strong feelings can decline. Some lawmakers propose going one step further and presuming that people want to donate their organs unless they opt out.

In the end, Sunstein argues, by normalizing organ donation as a routine medical practice instead of a rare, important, and gruesome event, the policy would short-circuit our fear reactions and nudge us toward a positive societal goal. It is this type of policy that Ropeik is trying to get the administration to think about, and it is the next step in risk perception and risk communication. "Our risk perception is flawed enough to create harm," he says, "but it's something society can do something about."

DAVID DOBBS

Beautiful Brains

FROM *National Geographic*

ALTHOUGH YOU KNOW your teenager takes some chances, it can be a shock to hear about them. One fine May morning not long ago my oldest son, seventeen at the time, phoned to tell me that he had just spent a couple hours at the state police barracks. Apparently he had been driving "a little fast." What, I asked, was "a little fast"? Turns out this product of my genes and loving care, the boy-man I had swaddled, coddled, cooed at, and then pushed and pulled to the brink of manhood, had been flying down the highway at 113 miles an hour.

"That's more than a little fast," I said.

He agreed. In fact, he sounded somber and contrite. He did not object when I told him he'd have to pay the fines and probably a lawyer. He did not argue when I pointed out that if anything happens at that speed—a dog in the road, a blown tire, a sneeze—he dies. He was in fact almost irritatingly reasonable. He even proffered that the cop did the right thing in stopping him, for, as he put it, "We can't all go around doing 113."

He did, however, object to one thing. He didn't like it that one of the several citations he had received was for reckless driving.

"Well," I huffed, sensing an opportunity to finally yell at him, "what would you call it?"

"It's just not accurate," he said calmly. "'Reckless' sounds like you're not paying attention. But I was. I made a deliberate point of doing this on an empty stretch of dry interstate, in broad daylight, with good sight lines and no traffic. I mean, I wasn't just gunning the thing. I was driving.

"I guess that's what I want you to know. If it makes you feel any better, I was really focused."

Actually, it did make me feel better. That bothered me, for I didn't understand why. Now I do.

My son's high-speed adventure raised the question long asked by people who have pondered the class of humans we call teenagers: What on earth was he doing? Parents often phrase this question more colorfully. Scientists put it more coolly. They ask, What can explain this behavior? But even that is just another way of wondering, What is wrong with these kids? Why do they act this way? The question passes judgment even as it inquires.

Through the ages, most answers have cited dark forces that uniquely affect the teen. Aristotle concluded more than 2,300 years ago that "the young are heated by Nature as drunken men by wine." A shepherd in William Shakespeare's *The Winter's Tale* wishes "there were no age between ten and three-and-twenty, or that youth would sleep out the rest; for there is nothing in the between but getting wenches with child, wronging the ancientry, stealing, fighting." His lament colors most modern scientific inquiries as well. G. Stanley Hall, who formalized adolescent studies with his 1904 *Adolescence: Its Psychology and Its Relations to Physiology, Anthropology, Sociology, Sex, Crime, Religion and Education*, believed this period of "storm and stress" replicated earlier, less civilized stages of human development. Freud saw adolescence as an expression of torturous psychosexual conflict; Erik Erikson, as the most tumultuous of life's several identity crises. Adolescence: always a problem.

Such thinking carried into the late twentieth century, when researchers developed brain-imaging technology that enabled them to see the teen brain in enough detail to track both its physical development and its patterns of activity. These imaging tools offered a new way to ask the same question—what's wrong with these kids? —and revealed an answer that surprised almost everyone. Our brains, it turned out, take much longer to develop than we had thought. This revelation suggested both a simplistic, unflattering explanation for teens' maddening behavior—and a more complex, affirmative explanation as well.

The first full series of scans of the developing adolescent brain —a National Institutes of Health (NIH) project that studied over a hundred young people as they grew up during the 1990s—showed that our brains undergo a massive reorganization between our twelfth and twenty-fifth years. The brain doesn't actually grow very much during this period. It has already reached 90 percent of its full size by the time a person is six, and a thickening skull accounts for most head growth afterward. But as we move through adolescence, the brain undergoes extensive remodeling, resembling a network and wiring upgrade.

For starters, the brain's axons—the long nerve fibers that neurons use to send signals to other neurons—become gradually more insulated with a fatty substance called myelin (the brain's white matter), eventually boosting the axons' transmission speed up to a hundred times. Meanwhile, dendrites, the branchlike extensions that neurons use to receive signals from nearby axons, grow twiggier, and the most heavily used synapses—the little chemical junctures across which axons and dendrites pass notes—grow richer and stronger. At the same time, synapses that see little use begin to wither. This synaptic pruning, as it is called, causes the brain's cortex—the outer layer of gray matter where we do much of our conscious and complicated thinking—to become thinner but more efficient. Taken together, these changes make the entire brain a much faster and more sophisticated organ.

This process of maturation, once thought to be largely finished by elementary school, continues throughout adolescence. Imaging work done since the 1990s shows that these physical changes move in a slow wave from the brain's rear to its front, from areas close to the brain stem that look after older and more behaviorally basic functions, such as vision, movement, and fundamental processing, to the evolutionarily newer and more complicated thinking areas up front. The corpus callosum, which connects the brain's left and right hemispheres and carries traffic essential to many advanced brain functions, steadily thickens. Stronger links also develop between the hippocampus, a sort of memory directory, and frontal areas that set goals and weigh different agendas; as a result, we get better at integrating memory and experience into our decisions. At the same time, the frontal areas develop greater speed and richer connections, allowing us to generate and weigh far more variables and agendas than before.

When this development proceeds normally, we get better at balancing impulse, desire, goals, self-interest, rules, ethics, and even altruism, generating behavior that is more complex and, sometimes at least, more sensible. But at times, and especially at first, the brain does this work clumsily. It's hard to get all those new cogs to mesh.

Beatriz Luna, a University of Pittsburgh professor of psychiatry who uses neuroimaging to study the teen brain, used a simple test that illustrates this learning curve. Luna scanned the brains of children, teens, and twentysomethings while they performed an antisaccade task, a sort of eyes-only video game where you have to stop yourself from looking at a suddenly appearing light. You view a screen on which the red crosshairs at the center occasionally disappear just as a light flickers elsewhere on the screen. Your instructions are to not look at the light and instead to look in the opposite direction. A sensor detects any eye movement. It's a tough assignment, since flickering lights naturally draw our attention. To succeed, you must override both a normal impulse to attend to new information and curiosity about something forbidden. Brain geeks call this response inhibition.

Ten-year-olds stink at it, failing about 45 percent of the time. Teens do much better. In fact, by age fifteen they can score as well as adults if they're motivated, resisting temptation about 70 to 80 percent of the time. What Luna found most interesting, however, was not those scores. It was the brain scans she took while people took the test. Compared with adults, teens tended to make less use of brain regions that monitor performance, spot errors, plan, and stay focused—areas the adults seemed to bring online automatically. This let the adults use a variety of brain resources and better resist temptation, while the teens used those areas less often and more readily gave in to the impulse to look at the flickering light —just as they're more likely to look away from the road to read a text message.

If offered an extra reward, however, teens showed they could push those executive regions to work harder, improving their scores. And by age twenty, their brains respond to this task much as the adults' do. Luna suspects the improvement comes as richer networks and faster connections make the executive region more effective.

These studies help explain why teens behave with such vexing inconsistency: beguiling at breakfast, disgusting at dinner; masterful on Monday, sleepwalking on Saturday. Along with lacking experience generally, they're still learning to use their brains' new networks. Stress, fatigue, or challenges can cause a misfire. Abigail Baird, a Vassar psychologist who studies teens, calls this "neural gawkiness"—an equivalent to the physical awkwardness teens sometimes display while mastering their growing bodies.

The slow and uneven developmental arc revealed by these imaging studies offers an alluringly pithy explanation for why teens may do stupid things like drive at 113 miles an hour, aggrieve their ancientry, and get someone (or get gotten) with child: they act that way because their brains aren't done! You can see it right there in the scans!

This view, as titles from the explosion of scientific papers and popular articles about the "teen brain" put it, presents adolescents as "works in progress" whose "immature brains" lead some to question whether they are in a state "akin to mental retardation."

The story you're reading right now, however, tells a different scientific tale about the teen brain. Over the past five years or so, even as the work-in-progress story spread into our culture, the discipline of adolescent brain studies learned to do some more complex thinking of its own. A few researchers began to view recent brain and genetic findings in a brighter, more flattering light, one distinctly colored by evolutionary theory. The resulting account of the adolescent brain—call it the adaptive-adolescent story—casts the teen less as a rough draft than as an exquisitely sensitive, highly adaptable creature wired almost perfectly for the job of moving from the safety of home into the complicated world outside.

This view will likely sit better with teens. More important, it sits better with biology's most fundamental principle, that of natural selection. Selection is hell on dysfunctional traits. If adolescence is essentially a collection of them—angst, idiocy, and haste; impulsiveness, selfishness, and reckless bumbling—then how did those traits survive selection? They couldn't—not if they were the period's most fundamental or consequential features.

The answer is that those troublesome traits don't really characterize adolescence; they're just what we notice most because they annoy us or put our children in danger. As B. J. Casey, a neuroscientist at Weill Cornell Medical College, who has spent nearly a decade applying brain and genetic studies to our understanding of adolescence, puts it, "We're so used to seeing adolescence as a problem. But the more we learn about what really makes this period unique, the more adolescence starts to seem like a highly functional, even adaptive period. It's exactly what you'd need to do the things you have to do then."

To see past the distracting, dopey teenager and glimpse the adaptive adolescent within, we should look not at specific, sometimes startling, behaviors, such as skateboarding down stairways or dating fast company, but at the broader traits that underlie those acts.

Let's start with the teen's love of the thrill. We all like new and exciting things, but we never value them more highly than we do during adolescence. Here we hit a high in what behavioral scientists call sensation seeking: the hunt for the neural buzz, the jolt of the unusual or unexpected.

Seeking sensation isn't necessarily impulsive. You might plan a sensation-seeking experience—a skydive or a fast drive—quite deliberately, as my son did. Impulsivity generally drops throughout life, starting at about age ten, but this love of the thrill peaks at around age fifteen. And although sensation seeking can lead to dangerous behaviors, it can also generate positive ones: the urge to meet more people, for instance, can create a wider circle of friends, which generally makes us healthier, happier, safer, and more successful.

This upside probably explains why an openness to the new, though it can sometimes kill the cat, remains a highlight of adolescent development. A love of novelty leads directly to useful experience. More broadly, the hunt for sensation provides the inspiration needed to "get you out of the house" and into new terrain, as Jay Giedd, a pioneering researcher in teen brain development at NIH, puts it.

Also peaking during adolescence (and perhaps aggrieving the ancientry the most) is risk taking. We court risk more avidly as teens than at any other time. This shows reliably in the lab, where teens take more chances in controlled experiments involving everything from card games to simulated driving. And it shows in real life, where the period from roughly fifteen to twenty-five brings peaks in all sorts of risky ventures and ugly outcomes. This age group dies of accidents of almost every sort (other than work accidents) at high rates. Most long-term drug or alcohol abuse starts during adolescence, and even people who later drink responsibly often drink too much as teens. Especially in cultures where teenage driving is common, this takes a gory toll: in the United States, one in three teen deaths is from car crashes, many involving alcohol.

Are these kids just being stupid? That's the conventional explanation: they're not thinking, or, by the work-in-progress model, their puny developing brains fail them.

Yet these explanations don't hold up. As Laurence Steinberg, a developmental psychologist specializing in adolescence at Temple University, points out, even fourteen- to seventeen-year-olds—the biggest risk takers—use the same basic cognitive strategies that adults do, and they usually reason their way through problems just as well as adults. Contrary to popular belief, they also fully recognize that they're mortal. And like adults, says Steinberg, "teens actually overestimate risk."

So if teens think as well as adults do and recognize risk just as well, why do they take more chances? Here, as elsewhere, the problem lies less in what teens lack compared with adults than in what they have more of. Teens take more risks not because they don't understand the dangers but because they weigh risk versus reward differently: in situations where risk can get them something they want, they value the reward more heavily than adults do.

A video game Steinberg uses draws this out nicely. In the game, you try to drive across town in as little time as possible. Along the way you encounter several traffic lights. As in real life, the traffic lights sometimes turn from green to yellow as you approach them, forcing a quick go-or-stop decision. You save time—and score more points—if you drive through before the light turns red. But if you try to drive through and don't beat the red, you lose even more time than you would have if you had stopped for it. Thus the game rewards you for taking a certain amount of risk but punishes you for taking too much.

When teens drive the course alone, in what Steinberg calls the emotionally "cool" situation of an empty room, they take risks at about the same rates as adults. Add stakes that the teen cares about, however, and the situation changes. In this case Steinberg added friends: when he brought a teen's friends into the room to watch, the teen would take twice as many risks, trying to gun it through lights he'd stopped for before. The adults, meanwhile, drove no differently with a friend watching.

To Steinberg, this shows clearly that risk taking rises not from puny thinking but from a higher regard for reward.

"They didn't take more chances because they suddenly downgraded the risk," says Steinberg. "They did so because they gave more weight to the payoff."

Researchers such as Steinberg and Casey believe this risk-friendly weighing of cost versus reward has been selected for because, over the course of human evolution, the willingness to take risks during this period of life has granted an adaptive edge. Succeeding often requires moving out of the home and into less secure situations. "The more you seek novelty and take risks," says Baird, "the better you do." This responsiveness to reward thus works like the desire for new sensation: it gets you out of the house and into new turf.

As Steinberg's driving game suggests, teens respond strongly to social rewards. Physiology and evolutionary theory alike offer explanations for this tendency. Physiologically, adolescence brings a peak in the brain's sensitivity to dopamine, a neurotransmitter that appears to prime and fire reward circuits and aids in learning patterns and making decisions. This helps explain the teen's quickness of learning and extraordinary receptivity to reward—and his keen, sometimes melodramatic reaction to success as well as defeat.

The teen brain is similarly attuned to oxytocin, another neural hormone, which (among other things) makes social connections in particular more rewarding. The neural networks and dynamics associated with general reward and social interactions overlap heavily. Engage one, and you often engage the other. Engage them during adolescence, and you light a fire.

This helps explain another trait that marks adolescence. Teens prefer the company of those their own age more than ever before or after. At one level, this passion for same-age peers merely expresses in the social realm the teen's general attraction to novelty: teens offer teens far more novelty than familiar old family does.

Yet teens gravitate toward peers for another, more powerful reason: to invest in the future rather than the past. We enter a world made by our parents. But we will live most of our lives and prosper (or not) in a world run and remade by our peers. Knowing, understanding, and building relationships with them bears critically on success. Socially savvy rats or monkeys, for instance, generally get the best nesting areas or territories, the most food and water, more allies, and more sex with better and fitter mates. And no species is more intricately and deeply social than humans are.

This supremely human characteristic makes peer relations not a sideshow but the main show. Some brain-scan studies, in fact, suggest that our brains react to peer exclusion much as they respond to threats to physical health or food supply. At a neural level, in other words, we perceive social rejection as a threat to existence. Knowing this might make it easier to abide the hysteria of a thirteen-year-old deceived by a friend or the gloom of a fifteen-year-old not invited to a party. These people! we lament. They react to social ups and downs as if their fates depended upon them! They're right. They do.

Excitement, novelty, risk, the company of peers. These traits may seem to add up to nothing more than doing foolish new stuff with friends. Look deeper, however, and you see that these traits that define adolescence make us more adaptive, both as individuals and as a species. That's doubtless why these traits, broadly defined, seem to show themselves in virtually all human cultures, modern or tribal. They may concentrate and express themselves more starkly in modern Western cultures, in which teens spend so much time with each other. But anthropologists have found that virtually all the world's cultures recognize adolescence as a distinct period in which adolescents prefer novelty, excitement, and peers. This near-universal recognition sinks the notion that it's a cultural construct.

Culture clearly shapes adolescence. It influences its expression and possibly its length. It can magnify its manifestations. Yet culture does not create adolescence. The period's uniqueness arises from genes and developmental processes that have been selected for over thousands of generations because they play an amplified role during this key transitional period: producing a creature optimally primed to leave a safe home and move into unfamiliar territory.

The move outward from home is the most difficult thing that humans do, as well as the most critical—not just for individuals but for a species that has shown an unmatched ability to master challenging new environments. In scientific terms, teenagers can be a pain in the ass. But they are quite possibly the most fully, crucially adaptive human beings around. Without them, humanity might not have spread so readily across the globe.

*

This adaptive-adolescence view, however accurate, can be tricky to come to terms with—the more so for parents dealing with teens in their most trying, contrary, or flat-out scary moments. It's reassuring to recast worrisome aspects as signs of an organism learning how to negotiate its surroundings. But natural selection swings a sharp edge, and the teen's sloppier moments can bring unbearable consequences. We may not run the risk of being killed in ritualistic battles or being eaten by leopards, but drugs, drinking, driving, and crime take a mighty toll. My son lives, and thrives, sans car, at college. Some of his high school friends, however, died during their driving experiments. Our children wield their adaptive plasticity amid small but horrific risks.

We parents, of course, often stumble too, as we try to walk the blurry line between helping and hindering our kids as they adapt to adulthood. The United States spends about $1 billion a year on programs to counsel adolescents on violence, gangs, suicide, sex, substance abuse, and other potential pitfalls. Few of them work.

Yet we can and do help. We can ward off some of the world's worst hazards and nudge adolescents toward appropriate responses to the rest. Studies show that when parents engage and guide their teens with a light but steady hand, staying connected but allowing independence, their kids generally do much better in life. Adolescents want to learn primarily, but not entirely, from their friends. At some level and at some times (and it's the parent's job to spot when), the teen recognizes that the parent can offer certain kernels of wisdom—knowledge valued not because it comes from parental authority but because it comes from the parent's own struggles to learn how the world turns. The teen rightly perceives that she must understand not just her parents' world but also the one she is entering. Yet if allowed to, she can appreciate that her parents once faced the same problems and may remember a few things worth knowing.

Meanwhile, in times of doubt, take inspiration from one last distinction of the teen brain—a final key to both its clumsiness and its remarkable adaptability. This is the prolonged plasticity of those late-developing frontal areas as they slowly mature. As noted earlier, these areas are the last to lay down the fatty myelin insulation—the brain's white matter—that speeds transmission. And at first glance this seems like bad news. If we need these areas for the complex task of entering the world, why aren't they running at full speed when the challenges are most daunting?

The answer is that speed comes at the price of flexibility. While a myelin coating greatly accelerates an axon's bandwidth, it also inhibits the growth of new branches from the axon. According to Douglas Fields, an NIH neuroscientist who has spent years studying myelin, "This makes the period when a brain area lays down myelin a sort of crucial period of learning—the wiring is getting upgraded, but once that's done, it's harder to change."

The window in which experience can best rewire those connections is highly specific to each brain area. Thus the brain's language centers acquire their insulation most heavily in the first thirteen years, when a child is learning language. The completed insulation consolidates those gains—but makes further gains, such as second languages, far harder to come by.

So it is with the forebrain's myelination during the late teens and early twenties. This delayed completion—a withholding of readiness—heightens flexibility just as we confront and enter the world that we will face as adults.

This long, slow, back-to-front developmental wave, completed only in the mid-twenties, appears to be a uniquely human adaptation. It may be one of our most consequential. It can seem a bit crazy that we humans don't wise up a bit earlier in life. But if we smartened up sooner, we'd end up dumber.

DAVID EAGLEMAN

The Brain on Trial

FROM *The Atlantic*

ON THE STEAMY first day of August 1966, Charles Whitman took an elevator to the top floor of the University of Texas Tower in Austin. The twenty-five-year-old climbed the stairs to the observation deck, lugging with him a footlocker full of guns and ammunition. At the top, he killed a receptionist with the butt of his rifle. Two families of tourists came up the stairwell; he shot them at point-blank range. Then he began to fire indiscriminately from the deck at people below. The first woman he shot was pregnant. As her boyfriend knelt to help her, Whitman shot him as well. He shot pedestrians in the street and an ambulance driver who came to rescue them.

The evening before, Whitman had sat at his typewriter and composed a suicide note:

> I don't really understand myself these days. I am supposed to be an average reasonable and intelligent young man. However, lately (I can't recall when it started) I have been a victim of many unusual and irrational thoughts.

By the time the police shot him dead, Whitman had killed thirteen people and wounded thirty-two more. The story of his rampage dominated national headlines the next day. And when police went to investigate his home for clues, the story became even stranger: in the early hours of the morning on the day of the shooting, he had murdered his mother and stabbed his wife to death in her sleep.

> It was after much thought that I decided to kill my wife, Kathy, tonight . . . I love her dearly, and she has been as fine a wife to me as any man could ever hope to have. I cannot rationa[l]ly pinpoint any specific reason for doing this . . .

Along with the shock of the murders lay another, more hidden, surprise: the juxtaposition of his aberrant actions with his unremarkable personal life. Whitman was an Eagle Scout and a former marine, studied architectural engineering at the University of Texas, and briefly worked as a bank teller and volunteered as a scoutmaster for Austin's Boy Scout Troop 5. As a child, he'd scored 138 on the Stanford-Binet IQ test, placing in the 99th percentile. So after his shooting spree from the University of Texas Tower, everyone wanted answers.

For that matter, so did Whitman. He requested in his suicide note that an autopsy be performed to determine if something had changed in his brain—because he suspected it had.

> I talked with a Doctor once for about two hours and tried to convey to him my fears that I felt [overcome by] overwhelming violent impulses. After one session I never saw the Doctor again, and since then I have been fighting my mental turmoil alone, and seemingly to no avail.

Whitman's body was taken to the morgue, his skull was put under the bone saw, and the medical examiner lifted the brain from its vault. He discovered that Whitman's brain harbored a tumor the diameter of a nickel. This tumor, called a glioblastoma, had blossomed from beneath a structure called the thalamus, impinged on the hypothalamus, and compressed a third region called the amygdala. The amygdala is involved in emotional regulation, especially of fear and aggression. By the late 1800s, researchers had discovered that damage to the amygdala caused emotional and social disturbances. In the 1930s, the researchers Heinrich Klüver and Paul Bucy demonstrated that damage to the amygdala in monkeys led to a constellation of symptoms, including lack of fear, blunting of emotion, and overreaction. Female monkeys with amygdala damage often neglected or physically abused their infants. In humans, activity in the amygdala increases when people are shown threatening faces, are put into frightening situations, or experience social phobias. Whitman's intuition about himself

—that something in his brain was changing his behavior—was spot-on.

Stories like Whitman's are not uncommon: legal cases involving brain damage crop up increasingly often. As we develop better technologies for probing the brain, we detect more problems and link them more easily to aberrant behavior. Take the 2000 case of a forty-year-old man we'll call Alex, whose sexual preferences suddenly began to transform. He developed an interest in child pornography—and not just a little interest but an overwhelming one. He poured his time into child-pornography web sites and magazines. He also solicited prostitution at a massage parlor, something he said he had never previously done. He reported later that he'd wanted to stop, but "the pleasure principle overrode" his restraint. He worked to hide his acts, but subtle sexual advances toward his prepubescent stepdaughter alarmed his wife, who soon discovered his collection of child pornography. He was removed from his house, found guilty of child molestation, and sentenced to rehabilitation in lieu of prison. In the rehabilitation program, he made inappropriate sexual advances toward the staff and other clients and was expelled and routed toward prison.

At the same time, Alex was complaining of worsening headaches. The night before he was to report for prison sentencing, he couldn't stand the pain anymore and took himself to the emergency room. He underwent a brain scan, which revealed a massive tumor in his orbitofrontal cortex. Neurosurgeons removed the tumor. Alex's sexual appetite returned to normal.

The year after the brain surgery, his pedophilic behavior began to return. The neuroradiologist discovered that a portion of the tumor had been missed in the surgery and was regrowing—and Alex went back under the knife. After the removal of the remaining tumor, his behavior again returned to normal.

When your biology changes, so can your decision making and your desires. The drives you take for granted ("I'm a heterosexual/homosexual," "I'm attracted to children/adults," "I'm aggressive/not aggressive," and so on) depend on the intricate details of your neural machinery. Although acting on such drives is popularly thought to be a free choice, the most cursory examination of the evidence demonstrates the limits of that assumption.

Alex's sudden pedophilia illustrates that hidden drives and desires can lurk undetected behind the neural machinery of socialization. When the frontal lobes are compromised, people become disinhibited, and startling behaviors can emerge. Disinhibition is commonly seen in patients with frontotemporal dementia, a tragic disease in which the frontal and temporal lobes degenerate. With the loss of that brain tissue, patients lose the ability to control their hidden impulses. To the frustration of their loved ones, these patients violate social norms in endless ways: shoplifting in front of store managers, removing their clothes in public, running stop signs, breaking out in song at inappropriate times, eating food scraps found in public trash cans, being physically aggressive or sexually transgressive. Patients with frontotemporal dementia commonly end up in courtrooms, where their lawyers, doctors, and embarrassed adult children must explain to the judge that the violation was not the perpetrator's *fault,* exactly: much of the brain has degenerated, and medicine offers no remedy. Fifty-seven percent of frontotemporal-dementia patients violate social norms, as compared with only 27 percent of Alzheimer's patients.

Changes in the balance of brain chemistry, even small ones, can also cause large and unexpected changes in behavior. Victims of Parkinson's disease offer an example. In 2001 families and caretakers of Parkinson's patients began to notice something strange. When patients were given a drug called pramipexole, some of them turned into gamblers. And not just casual gamblers but pathological gamblers. These were people who had never gambled much before, and now they were flying off to Vegas. One sixty-eight-year-old man amassed losses of more than $200,000 in six months at a series of casinos. Some patients became consumed with Internet poker, racking up unpayable credit-card bills. For several, the new addiction reached beyond gambling, to compulsive eating, excessive alcohol consumption, and hypersexuality.

What was going on? Parkinson's involves the loss of brain cells that produce a neurotransmitter known as dopamine. Pramipexole works by impersonating dopamine. But it turns out that dopamine is a chemical doing double duty in the brain. Along with its role in motor commands, it also mediates the reward systems, guiding a person toward food, drink, mates, and other things useful for survival. Because of dopamine's role in weighing the costs and benefits of decisions, imbalances in its levels can trigger gambling, overeating, and drug addiction—behaviors that result from a reward system gone awry. Physicians now watch for these behavioral changes as a possible side effect of drugs like pramipexole. Luckily, the negative effects of the drug are reversible—the physician simply lowers the dosage, and the compulsive gambling goes away.

The lesson from all these stories is the same: human behavior cannot be separated from human biology. If we like to believe that people make free choices about their behavior (as in "I don't gamble because I'm strong-willed"), cases like Alex the pedophile, the frontotemporal shoplifters, and the gambling Parkinson's patients may encourage us to examine our views more carefully. Perhaps not everyone is equally "free" to make socially appropriate choices.

Does the discovery of Charles Whitman's brain tumor modify your feelings about the senseless murders he committed? Does it affect the sentence you would find appropriate for him, had he survived that day? Does the tumor change the degree to which you consider the killings "his fault"? Couldn't you just as easily be unlucky enough to develop a tumor and lose control of your behavior?

On the other hand, wouldn't it be dangerous to conclude that people with a tumor are free of guilt and that they should be let off the hook for their crimes?

As our understanding of the human brain improves, juries are increasingly challenged with these sorts of questions. When a criminal stands in front of the judge's bench today, the legal system wants to know whether he is *blameworthy*. Was it his fault or his biology's fault?

I submit that this is the wrong question to be asking. The choices we make are inseparably yoked to our neural circuitry, and therefore we have no meaningful way to tease the two apart. The more we learn, the more the seemingly simple concept of blameworthiness becomes complicated, and the more the foundations of our legal system are strained.

If I seem to be heading in an uncomfortable direction—toward letting criminals off the hook—please read on, because I'm going to show the logic of a new argument, piece by piece. The upshot is that we can build a legal system more deeply informed by science, in which we will continue to take criminals off the streets, but we will customize sentencing, leverage new opportunities for rehabilitation, and structure better incentives for good behavior. Discoveries in neuroscience suggest a new way forward for law and order—one that will lead to a more cost-effective, humane, and flexible system than the one we have today. When modern brain science is laid out clearly, it is difficult to justify how our legal system can continue to function without taking what we've learned into account.

Many of us like to believe that all adults possess the same capacity to make sound choices. It's a charitable idea but demonstrably wrong. People's brains are vastly different.

Who you even have the possibility to be starts at conception. If you think genes don't affect how people behave, consider this fact: if you are a carrier of a particular set of genes, the probability that you will commit a violent crime is four times as high as it would be if you lacked those genes. You're three times as likely to commit robbery, five times as likely to commit aggravated assault, eight times as likely to be arrested for murder, and thirteen times as likely to be arrested for a sexual offense. The overwhelming majority of prisoners carry these genes; 98.1 percent of death-row inmates do. These statistics alone indicate that we cannot presume that everyone is coming to the table equally equipped in terms of drives and behaviors.

And this feeds into a larger lesson of biology: *we* are not the ones steering the boat of our behavior, at least not nearly as much as we believe. *Who we are* runs well below the surface of our conscious access, and the details reach back in time to before our birth, when the meeting of a sperm and an egg granted us certain attributes and not others. *Who we can be* starts with our molecular blueprints—a series of alien codes written in invisibly small strings of acids—well before we have anything to do with it. Each of us is, in part, a product of our inaccessible microscopic history. By the way, as regards that dangerous set of genes, you've probably heard of them. They are summarized as the Y chromosome. If you're a carrier, we call you a male.

Genes are part of the story, but they're not the whole story. We are likewise influenced by the environments in which we grow up. Substance abuse by a mother during pregnancy, maternal stress, and low birth weight all can influence how a baby will turn out as an adult. As a child grows, neglect, physical abuse, and head injury can impede mental development, as can the physical environment. (For example, the major public health movement to eliminate lead-based paint grew out of an understanding that ingesting lead can cause brain damage, making children less intelligent and, in some cases, more impulsive and aggressive.) And every experience throughout our lives can modify genetic expression—activating certain genes or switching others off—which in turn can inaugurate new behaviors. In this way, genes and environments intertwine.

When it comes to nature and nurture, the important point is that we choose neither one. We are each constructed from a genetic blueprint and then born into a world of circumstances that we cannot control in our most formative years. The complex interactions of genes and environment mean that all citizens—equal before the law—possess different perspectives, dissimilar personalities, and varied capacities for decision making. The unique patterns of neurobiology inside each of our heads cannot qualify as *choices;* these are the cards we're dealt.

Because we did not choose the factors that affected the formation and structure of our brain, the concepts of free will and personal responsibility begin to sprout question marks. Is it meaningful to say that Alex made bad *choices,* even though his brain tumor was not his fault? Is it justifiable to say that the patients with frontotemporal dementia or Parkinson's should be *punished* for their bad behavior?

It is problematic to imagine yourself in the shoes of someone breaking the law and conclude, "Well, *I* wouldn't have done that"—because if you weren't exposed to in utero cocaine, lead poisoning, and physical abuse, and he was, then you and he are not directly comparable. You cannot walk a mile in his shoes.

The legal system rests on the assumption that we are "practical reasoners," a term of art that presumes, at bottom, the existence of free will. The idea is that we use conscious deliberation when deciding how to act—that is, in the absence of external duress, we make free decisions. This concept of the practical reasoner is intuitive but problematic.

The existence of free will in human behavior is the subject of an ancient debate. Arguments in support of free will are typically based on direct subjective experience ("I *feel* like I made the decision to lift my finger just now"). But evaluating free will requires some nuance beyond our immediate intuitions.

Consider a decision to move or speak. It feels as though free will leads you to stick out your tongue, or scrunch up your face, or call someone a name. But free will is not *required* to play any role in these acts. People with Tourette's syndrome, for instance, suffer from involuntary movements and vocalizations. A typical Touretter may stick out his tongue, scrunch up his face, or call someone a name—all without *choosing* to do so.

We immediately learn two things from the Tourette's patient. First, actions can occur in the absence of free will. Second, the Tourette's patient has no *free won't*. He cannot use free will to override or control what subconscious parts of his brain have decided to do. What the lack of free will and the lack of free won't have in common is the lack of "free." Tourette's syndrome provides a case in which the underlying neural machinery does its thing, and we all agree that the person is not responsible.

This same phenomenon arises in people with a condition known as chorea, for whom actions of the hands, arms, legs, and face are involuntary, even though they certainly *look* voluntary: ask such a patient why she is moving her fingers up and down, and she will explain that she has no control over her hand. She cannot *not* do it. Similarly, some split-brain patients (who have had the two hemispheres of the brain surgically disconnected) develop alien-hand syndrome: while one hand buttons up a shirt, the other hand works to unbutton it. When one hand reaches for a pencil, the other bats it away. No matter how hard the patient tries, he cannot make his alien hand *not* do what it's doing. The movements are not "his" to freely start or stop.

Unconscious acts are not limited to unintended shouts or wayward hands; they can be surprisingly sophisticated. Consider Kenneth Parks, a twenty-three-year-old Canadian with a wife, a five-month-old daughter, and a close relationship with his in-laws (his mother-in-law described him as a "gentle giant"). Suffering from financial difficulties, marital problems, and a gambling addiction, he made plans to go see his in-laws to talk about his troubles.

In the wee hours of May 23, 1987, Kenneth arose from the couch on which he had fallen asleep, but he did not awaken. Sleepwalking, he climbed into his car and drove the fourteen miles to his in-laws' home. He broke in, stabbed his mother-in-law to death, and assaulted his father-in-law, who survived. Afterward, he drove himself to the police station. Once there, he said, "I think I have killed some people . . . . My hands," realizing for the first time that his own hands were severely cut.

Over the next year, Kenneth's testimony was remarkably consistent, even in the face of attempts to lead him astray: he remembered nothing of the incident. Moreover, while all parties agreed that Kenneth had undoubtedly committed the murder, they also agreed that he had no motive. His defense attorneys argued that this was a case of killing while sleepwalking, known as homicidal somnambulism.

Although critics cried "Faker!" sleepwalking is a verifiable phenomenon. On May 25, 1988, after lengthy consideration of electrical recordings from Kenneth's brain, the jury concluded that his actions had indeed been involuntary and declared him not guilty.

As with Tourette's sufferers, split-brain patients, and those with choreic movements, Kenneth's case illustrates that high-level behaviors can take place in the absence of free will. Like your heartbeat, breathing, blinking, and swallowing, even your mental machinery can run on autopilot. The crux of the question is whether *all* of your actions are fundamentally on autopilot or whether some little bit of you is "free" to choose, independent of the rules of biology.

This has always been the sticking point for philosophers and scientists alike. After all, there is no spot in the brain that is not densely interconnected with—and driven by—other brain parts. And that suggests that no part is independent and therefore "free." In modern science, it is difficult to find the gap into which to slip free will—the uncaused causer—because there seems to be no part of the machinery that does not follow in a causal relationship from the other parts.

Free will *may* exist (it may simply be beyond our current science), but one thing seems clear: if free will *does* exist, it has little room in which to operate. It can at best be a small factor riding on top of vast neural networks shaped by genes and environment. In fact, free will may end up being so small that we eventually think about bad decision making in the same way we think about any physical process, such as diabetes or lung disease.

The study of brains and behaviors is in the midst of a conceptual shift. Historically, clinicians and lawyers have agreed on an intuitive distinction between neurological disorders ("brain problems") and psychiatric disorders ("mind problems"). As recently as a century ago, a common approach was to get psychiatric patients to "toughen up," through deprivation, pleading, or torture. Not surprisingly, this approach was medically fruitless. After all, while psychiatric disorders tend to be the product of more subtle forms of brain pathology, they, too, are based in the biological details of the brain.

What accounts for the shift from blame to biology? Perhaps the largest driving force is the effectiveness of pharmaceutical treatments. No amount of threatening will chase away depression, but a little pill called fluoxetine often does the trick. Schizophrenic symptoms cannot be overcome by exorcism, but they can be controlled by risperidone. Mania responds not to talk or to ostracism but to lithium. These successes, most of them introduced in the past sixty years, have underscored the idea that calling some disorders "brain problems" while consigning others to the ineffable realm of "the psychic" does not make sense. Instead, we have begun to approach mental problems in the same way we might approach a broken leg. The neuroscientist Robert Sapolsky invites us to contemplate this conceptual shift with a series of questions:

> Is a loved one, sunk in a depression so severe that she cannot function, a case of a disease whose biochemical basis is as "real" as is the biochemistry of, say, diabetes, or is she merely indulging herself? Is a child doing poorly at school because he is unmotivated and slow, or because there is a neurobiologically based learning disability? Is a friend, edging towards a serious problem with substance abuse, displaying a simple lack of discipline, or suffering from problems with the neurochemistry of reward?

Acts cannot be understood separately from the biology of the actors—and this recognition has legal implications. Tom Bingham, Britain's former senior law lord, once put it this way:

> In the past, the law has tended to base its approach . . . on a series of rather crude working assumptions: adults of competent mental capacity are free to choose whether they will act in one way or another; they are presumed to act rationally, and in what they conceive to be their own best interests; they are credited with such foresight of the consequences of their actions as reasonable people in their position could ordinarily be expected to have; they are generally taken to mean what they say.
>
> Whatever the merits or demerits of working assumptions such as these in the ordinary range of cases, it is evident that they do not provide a uniformly accurate guide to human behaviour.

The more we discover about the circuitry of the brain, the more we tip away from accusations of indulgence, lack of motivation, and poor discipline—and toward the details of biology. The shift from blame to science reflects our modern understanding that our perceptions and behaviors are steered by deeply embedded neural programs.

Imagine a spectrum of culpability. On one end, we find people like Alex the pedophile or a patient with frontotemporal dementia who exposes himself in public. In the eyes of the judge and jury, these are people who suffered brain damage at the hands of fate and did not choose their neural situation. On the other end of the spectrum—the blameworthy side of the "fault" line—we find the common criminal, whose brain receives little study and about whom our current technology might be able to say little anyway. The overwhelming majority of lawbreakers are on this side of the line, because they don't have any obvious, measurable biological problems. They are simply thought of as freely choosing actors.

Such a spectrum captures the common intuition that juries hold regarding blameworthiness. But there is a deep problem with this intuition. Technology will continue to improve, and as we grow better at measuring problems in the brain, the fault line will drift into the territory of people we currently hold fully accountable for their crimes. Problems that are now opaque will open up to examination by new techniques, and we may someday find that many types of bad behavior have a basic biological explanation—as has happened with schizophrenia, epilepsy, depression, and mania.

Today, neuroimaging is a crude technology, unable to explain the details of individual behavior. We can detect only large-scale problems, but in the coming decades, we will be able to detect patterns at unimaginably small levels of the microcircuitry that correlate with behavioral problems. Neuroscience will be better able to say why people are predisposed to act the way they do. As we become more skilled at specifying how behavior results from the microscopic details of the brain, more defense lawyers will point to biological mitigators of guilt, and more juries will place defendants on the not-blameworthy side of the line.

This puts us in a strange situation. After all, a just legal system cannot define culpability simply by the limitations of current technology. Expert medical testimony generally reflects only whether we yet have names and measurements for a problem, not whether a problem exists. A legal system that declares a person culpable at the beginning of a decade and not culpable at the end is one in which culpability carries no clear meaning.

The crux of the problem is that it no longer makes sense to ask, "To what extent was it his *biology,* and to what extent was it *him?*" because we now understand that there is no meaningful distinction between a person's biology and his decision making. They are inseparable.

While our current style of punishment rests on a bedrock of personal volition and blame, our modern understanding of the brain suggests a different approach. Blameworthiness should be removed from the legal argot. It is a backward-looking concept that demands the impossible task of untangling the hopelessly complex web of genetics and environment that constructs the trajectory of a human life.

Instead of debating culpability, we should focus on what to do, *moving forward,* with an accused lawbreaker. I suggest that the legal system *has* to become forward-looking, primarily because it can no longer hope to do otherwise. As science complicates the question of culpability, our legal and social policy will need to shift toward a different set of questions: How is a person likely to behave in the future? Are criminal actions likely to be repeated? Can this person be helped toward prosocial behavior? How can incentives be realistically structured to deter crime?

The important change will be in the *way* we respond to the vast range of criminal acts. Biological explanation will not exculpate criminals; we will still remove from the streets lawbreakers who prove overaggressive, underempathetic, and poor at controlling their impulses. Consider, for example, that the majority of known serial killers were abused as children. Does this make them less blameworthy? Who cares? It's the wrong question. The knowledge that they were abused encourages us to support social programs to prevent child abuse, but it does nothing to change the way we deal with the particular serial murderer standing in front of the bench. We still need to keep him off the streets, irrespective of his past misfortunes. The child abuse cannot serve as an excuse to let him go; the judge must keep society safe.

Those who break social contracts need to be confined, but in this framework, the future is more important than the past. Deeper biological insight into behavior will foster a better understanding of recidivism—and this offers a basis for empirically based sentencing. Some people will need to be taken off the streets for a longer time (even a lifetime), because their likelihood of reoffense is high; others, because of differences in neural constitution, are less likely to recidivate and so can be released sooner.

The law is already forward-looking in some respects: consider the leniency afforded a crime of passion versus a premeditated murder. Those who commit the former are less likely to recidivate than those who commit the latter, and their sentences sensibly reflect that. Likewise, American law draws a bright line between criminal acts committed by minors and those by adults, punishing the latter more harshly. This approach may be crude, but the intuition behind it is sound: adolescents command lesser skills in decision making and impulse control than do adults; a teenager's brain is simply not like an adult's brain. Lighter sentences are appropriate for those whose impulse control is likely to improve naturally as adolescence gives way to adulthood.

Taking a more scientific approach to sentencing, case by case, could move us beyond these limited examples. For instance, important changes are happening in the sentencing of sex offenders. In the past, researchers asked psychiatrists and parole-board members how likely specific sex offenders were to relapse when let out of prison. Both groups had experience with sex offenders, so predicting who would go straight and who would come back seemed simple. But surprisingly, the expert guesses showed almost no correlation with the actual outcomes. The psychiatrists and parole-board members had only slightly better predictive accuracy than coin-flippers. This astounded the legal community.

So researchers tried a more actuarial approach. They set about recording dozens of characteristics of some 23,000 released sex offenders: whether the offender had unstable employment, had been sexually abused as a child, was addicted to drugs, showed remorse, had deviant sexual interests, and so on. Researchers then tracked the offenders for an average of five years after release to see who wound up back in prison. At the end of the study, they computed which factors best explained the reoffense rates, and from these and later data they were able to build actuarial tables to be used in sentencing.

Which factors mattered? Take, for instance, low remorse, denial of the crime, and sexual abuse as a child. You might guess that these factors would correlate with sex offenders' recidivism. But you would be wrong: those factors offer no predictive power. How about antisocial personality disorder and failure to complete treatment? These offer somewhat more predictive power. But among the strongest predictors of recidivism are prior sexual offenses and sexual interest in children. When you compare the predictive power of the actuarial approach with that of the parole boards and psychiatrists, there is no contest: numbers beat intuition. In courtrooms across the nation, these actuarial tests are now used in presentencing to modulate the length of prison terms.

We will never know with certainty what someone will do upon release from prison, because real life is complicated. But greater predictive power is hidden in the numbers than people generally expect. Statistically based sentencing is imperfect, but it nonetheless allows evidence to trump folk intuition, and it offers customization in place of the blunt guidelines that the legal system typically employs. The current actuarial approaches do not require a deep understanding of genes or brain chemistry, but as we introduce more science into these measures—for example, with neuroimaging studies—the predictive power will only improve. (To make such a system immune to government abuse, the data and equations that compose the sentencing guidelines must be transparent and available online for anyone to verify.)

*

Beyond customized sentencing, a forward-thinking legal system informed by scientific insights into the brain will enable us to stop treating prison as a one-size-fits-all solution. To be clear, I'm not opposed to incarceration, and its purpose is not limited to the removal of dangerous people from the streets. The prospect of incarceration deters many crimes, and time actually spent in prison can steer some people away from further criminal acts upon their release. But that works only for those whose brains function normally. The problem is that prisons have become our de facto mental-health-care institutions—and inflicting punishment on the mentally ill usually has little influence on their future behavior. An encouraging trend is the establishment of mental-health courts around the nation: through such courts, people with mental illnesses can be helped while confined in a tailored environment. Cities such as Richmond, Virginia, are moving in this direction, for reasons of justice as well as cost-effectiveness. Sheriff C. T. Woody, who estimates that nearly 20 percent of Richmond's prisoners are mentally ill, told *CBS News*, "The jail isn't a place for them. They should be in a mental-health facility." Similarly, many jurisdictions are opening drug courts and developing alternative sentences; they have realized that prisons are not as useful for solving addictions as are meaningful drug-rehabilitation programs.

A forward-thinking legal system will also parlay biological understanding into customized rehabilitation, viewing criminal behavior the way we understand other medical conditions, such as epilepsy, schizophrenia, and depression—conditions that now allow the seeking and giving of help. These and other brain disorders find themselves on the not-blameworthy side of the fault line, where they are now recognized as biological, not demonic, issues.

Many people recognize the long-term cost-effectiveness of rehabilitating offenders instead of packing them into overcrowded prisons. The challenge has been the dearth of new ideas about *how* to rehabilitate them. A better understanding of the brain offers new ideas. For example, poor impulse control is characteristic of many prisoners. These people generally can express the difference between right and wrong actions, and they understand the disadvantages of punishment—but they are handicapped by poor control of their impulses. Whether as a result of anger or temptation, their actions override reasoned consideration of the future.

If it seems difficult to empathize with people who have poor impulse control, just think of all the things you succumb to against your better judgment. Alcohol? Chocolate cake? Television? It's not that we don't know what's best for us, it's simply that the frontal-lobe circuits representing long-term considerations can't always win against short-term desire when temptation is in front of us.

With this understanding in mind, we can modify the justice system in several ways. One approach, advocated by Mark A. R. Kleiman, a professor of public policy at the University of California, Los Angeles, is to ramp up the certainty and swiftness of punishment—for instance, by requiring drug offenders to undergo twice-weekly drug testing, with automatic, immediate consequences for failure—thereby not relying on distant abstraction alone. Similarly, economists have suggested that the drop in crime since the early 1990s has been due, in part, to the increased presence of police on the streets: their visibility shores up support for the parts of the brain that weigh long-term consequences.

We may be on the cusp of finding new rehabilitative strategies as well, affording people better control of their behavior even in the absence of external authority. To help a citizen reintegrate into society, the ethical goal is to change him *as little as possible* while bringing his behavior into line with society's needs. My colleagues and I are proposing a new approach, one that grows from the understanding that the brain operates like a team of rivals, with different neural populations competing to control the single output channel of behavior. Because it's a competition, the outcome can be tipped. I call the approach "the prefrontal workout."

The basic idea is to give the frontal lobes practice in squelching the short-term brain circuits. To this end, my colleagues Stephen LaConte and Pearl Chiu have begun providing real-time feedback to people during brain scanning. Imagine that you'd like to quit smoking cigarettes. In this experiment, you look at pictures of cigarettes during brain imaging, and the experimenters measure which regions of your brain are involved in the craving. Then they show you the activity in those networks, represented by a vertical bar on a computer screen, while you look at more cigarette pictures. The bar acts as a thermometer for your craving: if your craving networks are revving high, the bar is high; if you're suppressing your craving, the bar is low. Your job is to make the bar go down. Perhaps you have insight into what you're doing to resist the craving; perhaps the mechanism is inaccessible. In any case, you try out different mental avenues until the bar begins to slowly sink. When it goes all the way down, that means you've successfully recruited frontal circuitry to squelch the activity in the networks involved in impulsive craving. The goal is for the long term to trump the short term. Still looking at pictures of cigarettes, you practice making the bar go down over and over, until you've strengthened those frontal circuits. By this method, you're able to visualize the activity in the parts of your brain that need modulation, and you can witness the effects of different mental approaches you might take.

If this sounds like biofeedback from the 1970s, it is—but this time with vastly more sophistication, monitoring specific networks inside the head rather than a single electrode on the skin. This research is just beginning, so the method's efficacy is not yet known—but if it works well, it will be a game changer. We will be able to take it to the incarcerated population, especially those approaching release, to try to help them avoid coming back through the revolving prison doors.

This prefrontal workout is designed to better balance the debate between the long- and short-term parts of the brain, giving the option of reflection before action to those who lack it. And really, that's all maturation is. The main difference between teenage and adult brains is the development of the frontal lobes. The human prefrontal cortex does not fully develop until the early twenties, and this fact underlies the impulsive behavior of teenagers. The frontal lobes are sometimes called the organ of socialization, because becoming socialized largely involves developing the circuitry to squelch our first impulses.

This explains why damage to the frontal lobes unmasks unsocialized behavior that we would never have thought was hidden inside us. Recall the patients with frontotemporal dementia who shoplift, expose themselves, and burst into song at inappropriate times. The networks for those behaviors have been lurking under the surface all along, but they've been masked by normally functioning frontal lobes. The same sort of unmasking happens in people who go out and get rip-roaring drunk on a Saturday night: they're disinhibiting normal frontal-lobe function and letting more impulsive networks climb onto the main stage. After training at the prefrontal gym, a person might still crave a cigarette, but he'll know how to beat the craving instead of letting it win. It's not that we don't want to enjoy our impulsive thoughts *(Mmm, cake)*, it's merely that we want to endow the frontal cortex with some control over whether we act upon them *(I'll pass)*. Similarly, if a person thinks about committing a criminal act, that's permissible as long as he doesn't take action.

For the pedophile, we cannot hope to control whether he is attracted to children. That he never acts on the attraction may be the best we can hope for, especially as a society that respects individual rights and freedom of thought. Social policy can hope only to prevent impulsive thoughts from tipping into behavior without reflection. The goal is to give more control to the neural populations that care about long-term consequences—to inhibit impulsivity, to encourage reflection. If a person thinks about long-term consequences and still decides to move forward with an illegal act, then we'll respond accordingly. The prefrontal workout leaves the brain intact—no drugs or surgery—and uses the natural mechanisms of brain plasticity to help the brain help itself. It's a tune-up rather than a product recall.

We have hope that this approach represents the correct model: it is grounded simultaneously in biology and in libertarian ethics, allowing a person to help himself by improving his long-term decision making. Like any scientific attempt, it could fail for any number of unforeseen reasons. But at least we have reached a point where we can develop new ideas rather than assuming that repeated incarceration is the single practical solution for deterring crime.

Along any axis that we use to measure human beings, we discover a wide-ranging distribution, whether in empathy, intelligence, impulse control, or aggression. People are not created equal. Although this variability is often imagined to be best swept under the rug, it is in fact the engine of evolution. In each generation, nature tries out as many varieties as it can produce, along all available dimensions.

Variation gives rise to lushly diverse societies—but it serves as a source of trouble for the legal system, which is largely built on the premise that humans are all equal before the law. This myth of human equality suggests that people are equally capable of controlling impulses, making decisions, and comprehending consequences. While admirable in spirit, the notion of neural equality is simply not true.

As brain science improves, we will better understand that people exist along continua of capabilities rather than in simplistic categories. And we will be better able to tailor sentencing and rehabilitation for the individual rather than maintain the pretense that all brains respond identically to complex challenges and that all people therefore deserve the same punishments. Some people wonder whether it's unfair to take a scientific approach to sentencing—after all, where's the humanity in that? But what's the alternative? As it stands now, ugly people receive longer sentences than attractive people; psychiatrists have no capacity to guess which sex offenders will reoffend; and our prisons are overcrowded with drug addicts and the mentally ill, both of whom could be better helped by rehabilitation. So is current sentencing really superior to a scientifically informed approach?

Neuroscience is beginning to touch on questions that were once only in the domain of philosophers and psychologists, questions about how people make decisions and the degree to which those decisions are truly "free." These are not idle questions. Ultimately, they will shape the future of legal theory and create a more biologically informed jurisprudence.

PART FIVE

Society and Environment

JOHN SEABROOK

Crush Point

FROM *The New Yorker*

ON THANKSGIVING DAY, 2008, shoppers began lining up outside the Walmart in Valley Stream, Long Island, at 5:30 P.M., near a small, handwritten sign that read BLITZ LINE STARTS HERE. Like many other retailers holding "doorbuster" Black Friday sales, Walmart was offering deep discounts on a limited number of TVs, iPods, DVD players, and other coveted products. Only two months earlier, the US economy had nearly collapsed, and although the Christmas shopping season was looking dismal, there was still some dim hope that the nation might be able to shop its way out of disaster, as we were advised to do after 9/11.

By two in the morning, the line ran the length of the building, passed Petland, turned at a wire fence, and stretched far into the bleak parking lot of the Green Acres Mall, a tundra of frosted tarmac. There were already more than a thousand people. Store managers had placed eight interlocking plastic barriers between the front of the line and the outer doors to the store to create a buffer zone that would keep people from crowding around the entrance. But at three, people began jumping the barriers. The store's assistant manager, Mike Sicuranza, spoke to the manager, Steve Sooknanan, who had gone to a hotel to rest, and told him that customers had breached the buffer zone. Sicuranza sounded frightened. Sooknanan told him to call the police.

The Nassau County police arrived soon after the call and, using bullhorns, ordered everyone to get back behind the barriers. The police were still there at four, when Sooknanan returned to the store. Shortly afterward, a Walmart employee brought some family members inside the barriers, angering the crowd. About two hundred shoppers pushed into the buffer zone. Those in front were squeezed against two sliding-glass outer doors that led into a glassed-in, high-ceilinged entrance vestibule that also held some vending machines. These had been pushed to the center of the space to prevent people from crossing it diagonally and entering through the exit doors. As more people gathered, in anticipation of the store's opening, at five A.M., the pressure on the doors built and they began to shake. "Push the doors in!" some chanted from the back.

Employees asked the police for help. According to a court filing, the police responded that dealing with this crowd was "not in their job description," and they left. Of the two-man security force that Walmart had hired for Blitz Day, only one had shown up, and he was inside the store. Shortly before five, the crowd had grown to about two thousand people. The store's asset-protection manager, Sal D'Amico, advised Sooknanan not to open the doors, but Sooknanan overruled him. He instructed eight to ten of his largest employees, most of whom worked in the stockroom, to stand at the sides of the vestibule as the outer doors were opened, and be ready to help anyone who tripped or fell.

One of those men was Jdimytai Damour, who lived in Jamaica, Queens; his parents were Haitian immigrants. Damour was thirty-four and beefy—at 6 feet 5 inches tall, he weighed around 480 pounds. Friends called him Jdidread, because he wore his hair in dreadlocks. He had been working at Walmart for about a week as a temporary employee in the stockroom. Like the others in the vestibule, he had no training in security or crowd control. A coworker had reportedly heard him say earlier, "I don't want to be here."

Just before five, the workers realized that a pregnant woman, Leana Lockley, a twenty-eight-year-old part-time college student from South Ozone Park, was being crushed against the glass on the outer doors. The managers slid them open just enough to pull Lockley inside the vestibule. The crowd surged forward, thinking that the store was opening. The workers shut the doors again and braced both sliding doors with their bodies to keep them from caving in, as Sooknanan initiated the festive countdown, a Walmart Blitz Day tradition. Ten, nine, eight . . . At zero, the doors were opened again. There was a loud cracking sound as both sliding doors burst from their frame, and the crowd boiled in.

Dennis Fitch, one of the workers standing at the entrance, was blown backward through the inner vestibule doors and into the store. Others managed to jump to safety atop the vending machines. Some attempted to form a human chain on the other side of the vestibule, to slow down the crowd rushing into the store. A crush soon developed inside the vestibule, but the people who were still outside, pushing forward, weren't aware of it. Leana Lockley was carried through the vestibule and into the store by the surge, and she tripped over an older woman, who was on the ground. As she got to her knees, she later said, she saw Damour next to her. "I was screaming that I was pregnant, I am sure he heard that," she told *Newsday*. "He was trying to block the people from pushing me down to the ground and trampling me. . . . It was a split second, and we had eye contact as we knew we were going to die."

Coworkers later testified that Damour was hit by one of the two sliding glass doors. As he went down, the door fell on top of him, and people fell over it. Maybe he got up again to help Lockley, but that's not clear in camera and cell-phone video footage of the scene. He just vanishes into the frantic tangle of limbs.

Lockley's husband, Shawn, was able to pull her out, badly bruised. Their baby, a healthy girl, was born the following April. But though "Big guy down" was broadcast over the walkie-talkies that some of Damour's coworkers carried, they had to fight their way through the crowd to reach him, and when they got there Damour's tongue was out and his eyes had rolled back. The cops arrived at 5:05 A.M. and performed CPR (a cell-phone video made its way to YouTube), without success. Damour was pronounced dead at Franklin Hospital Medical Center, in Valley Stream, at 6:03 A.M. The coroner's report did not mention any bruises, fractures, or internal injuries, as it would have if he'd been trampled to death; the cause of death was listed as asphyxia.

Crowds are a condition of urban life. On subways and sidewalks, in elevators and stores, we pass in and out of them in the course of a day without pausing to consider by what mechanisms our brains guide us through so easily, rarely touching so much as a stranger's shoulder. Crowds are often viewed as a necessary inconvenience of city living, but there are occasions when we gladly join them, pressing together at raves and rock concerts, at sporting events, victory parades, and big sales. Elias Canetti, in his 1960 book *Crowds and Power,* sees these times of physical communion with strangers as essential to transcending the fear of being touched. "The more fiercely people press together," he writes, "the more certain they feel that they do not fear each other." In fact, a crowd is most dangerous when density is greatest. The transition from fraternal smooshing to suffocating pressure—a "crowd crush"—often occurs almost imperceptibly; one doesn't realize what's happening until it's too late to escape. Something interrupts the flow of pedestrians—a blocked exit, say, while an escalator continues to feed people into a closed-off space, or a storm that causes everyone to start running for shelter at the same time. (In Belarus in 1999, fifty-two people died when a crowd tried to enter an underground railway station to keep dry.) At a certain point, you feel pressure on all sides of your body and realize that you can't raise your arms. You are pulled off your feet and welded into a block of people. The crowd force squeezes the air out of your lungs, and you struggle to take another breath.

John Fruin, a retired research engineer with the Port Authority of New York and New Jersey, is one of the founders of crowd studies in the United States. In a 1993 paper, "The Causes and Prevention of Crowd Disasters," he wrote, "At occupancies of about 7 persons per square meter the crowd becomes almost a fluid mass. Shock waves can be propagated through the mass sufficient to lift people off of their feet and propel them distances of 3 m (10 ft) or more. People may be literally lifted out of their shoes, and have clothing torn off. Intense crowd pressures, exacerbated by anxiety, make it difficult to breathe." Some people die standing up; others die in the pileup that follows a "crowd collapse," when someone goes down and more people fall over him. "Compressional asphyxia" is usually given as the cause of death in these circumstances.

Crowd disasters occur all over the world, and for a variety of reasons. According to a recent paper published in the journal *Disaster Medicine and Public Health Preparedness,* reports of human stampedes have more than doubled in each of the past two decades. In the developing world, they often occur at religious festivals. In November, hundreds of people died in Cambodia, in a crush that occurred on a bridge in Phnom Penh during the annual water festival; there were reports that the police had fired water cannons at people on the crowded bridge. Thousands have died making pilgrimages to Mecca in the past twenty years, mainly in the ritual called the Stoning of the Devil, which occurs near the Jamarat Bridge; 360 pilgrims were killed there in 2006. In India last month, more than a hundred Hindu worshipers died in a crush in the state of Kerala.

In the developed world, soccer games and rock concerts are the most likely events to generate deadly crowds. In 1989 in Sheffield, England, ninety-five people died after they were caught in a crowd crush at Hillsborough Stadium when fans were trying to get into a soccer match between Nottingham Forest and Liverpool. (A ninety-sixth victim was taken off life support four years later.) At a rock festival in Roskilde, Denmark, in 2000, nine people died after a crowd collapse that occurred near the stage while Pearl Jam was performing in front of an audience of 50,000. Last July twenty-one people were killed at the Love Parade, a free electronic-music festival in Duisburg, Germany, when a crush developed in a disused rail tunnel that led to the festival grounds. With the world's population increasing, and with more people moving to cities, crowds will become ever larger, and disasters more frequent, unless scientists and safety engineers can figure out how to prevent them from happening.

In the literature on crowd disasters, there is a striking incongruity between the way these events are depicted in the press and how they actually occur. In popular accounts, they are almost invariably described as "panics." The crowd is portrayed as a single, unified entity that acts according to "mob psychology"—a set of primitive instincts (fear, followed by flight) that favor self-preservation over the welfare of others and cause "stampedes" and "tramplings." But most crowd disasters are caused by "crazes"—people are usually moving toward something they want rather than away from something they fear, and if you're caught up in a crush, you're just as likely to die on your feet as under the feet of others, squashed by the pressure of bodies smashing into you. (Investigators collecting evidence in the aftermath of crowd disasters have found steel guardrails capable of withstanding a thousand pounds of pressure bent by crowd force.) In disasters not involving fire, panic is rarely the cause of fatalities, and even when fire is involved, as in the 1977 Beverly Hills Supper Club fire in Southgate, Kentucky, research has shown that people continue to help one another, even at the cost of their own lives.

So why do we still think in terms of panics and stampedes? In many crowd disasters, particularly those in the West where commercial interests are involved, different stakeholders are potentially responsible, including the organizers of the event, the venue owners and designers, and the public officials and private security firms whose job is to ensure crowd safety. In the aftermath of disasters, they all vigorously defend their interests, and rarely are any of them held accountable. But almost no one speaks for the crowd, and the crowd usually takes the blame.

The origins of the term *Black Friday* are obscure. Some think that it was first used by the police in Philadelphia to describe the snarled traffic and sidewalk hassles that came with the day after Thanksgiving and crowds arriving for the city's annual Army-Navy game. Others have defined Black Friday as the day that merchants' balance sheets cross over into the black. Either way, it is now a de facto national shopping holiday. On TV, images of people racing through the aisles of stores for sale-priced items, in a sort of American Pamplona, have become as much a part of the day after Thanksgiving as leftovers. Shoppers get discounts, programmers get some lively content for a slow news day, and retailers get free publicity: a good deal for everyone, except for the clerks who have to work that day, breaking up fights among shoppers and cleaning up the mess left behind.

There had been injuries on previous Black Fridays, but no one had ever died before Jdimytai Damour went down in the Valley Stream Walmart. His death, and the "Walmart Stampede" that caused it, was the lead story on news channels across the country that evening, and it provoked a vast outcry of horror. In days of commentary that followed, the crowd was widely vilified. The tone of much of the reaction was captured by a letter writer to the New York *Post,* who blamed "the animals (you know who you are) who stampeded that poor man at Walmart on Black Friday: You are a perfect example of the depraved decadence of society today."

A Walmart senior vice president, Hank Mullany, said in a statement, "Our thoughts and prayers go out to the family of the deceased. We are continuing to work closely with local law enforcement, and we are reaching out to those involved." Investigators would be reviewing video collected from security cameras and looking at purchases made with credit cards in an effort to identify individuals who may have witnessed or been involved in Damour's death. But even if investigators could pick out, amid the flailing limbs and hurtling bodies in the videos, those who had harmed Damour, who could say that he or she wasn't pushed by the person behind? And at any rate, the police seemed to be in no mood to "work closely" with Walmart. Rather, they went out of their way to blame Walmart for the incident. Detective Lieutenant Michael Fleming, who was in charge of the investigation into Damour's death, said at the time, "I've heard other people call this an accident, but it is not. Certainly it was a foreseeable act."

Through the winter and spring of 2009, the Nassau County District Attorney's Office prepared to bring criminal charges against Walmart for felony reckless endangerment and misdemeanor reckless endangerment. The family of Jdimytai Damour filed a wrongful-death claim against the company. However, in early May 2009, the county's district attorney, Kathleen Rice, announced that her office had worked out a deal with Walmart that allowed it to avoid criminal charges. The company agreed to donate $1.5 million to various community projects and to create a $400,000 victims' fund. Walmart also agreed to implement a "crowd management" plan for future post–Thanksgiving Day events at each of its ninety-two New York stores. In return, Walmart would face no charges or criminal liability for the death of Jdimytai Damour. If the company failed to meet the standards set by an independent monitor for three years, the criminal case would be reinstated. Rice, noting that the maximum penalty Walmart would have faced was a $10,000 fine, said, "This agreement does more than any criminal prosecution could ever accomplish." Damour's father, Ogera Charles, saw things differently. "It's like if they were driving a car and they hit someone, killed him, and then just walked away," he told *Newsday*.

At the end of May, the Occupational Safety and Health Administration cited Walmart for committing a "serious violation" of the General Duty Clause of the OSHA Act. The clause states that an employer must furnish workers with a place of employment that is "free from recognized hazards that are causing or are likely to cause death or serious physical harm to his employees." In its complaint, OSHA listed the hazards that Damour and his coworkers faced there as "asphyxiation or being struck due to crowd crush, crowd surge or crowd trampling." The complaint also said that Walmart "did not use appropriate crowd management techniques to safely manage a large crowd of approximately 2,000 customers."

The proposed penalty was $7,000—not an enormous burden for the world's biggest retailer, which had total sales of $405 billion in 2010. But Walmart elected to contest the citation and hired the Washington, DC, law firm of Gibson, Dunn & Crutcher to handle the litigation. Walmart objected on multiple grounds. First, if crowd crushes and surges were recognized hazards, then why hadn't a single OSHA General Duty Clause citation ever referred to the dangers posed by crowds before? Walmart also maintained that it had taken steps to protect its workers from the crowd, but it could not have protected workers from this particular crowd. And finally, the violence caused by the crowd was a police issue and therefore beyond OSHA's jurisdiction.

A federal administrative-law judge, Covette Rooney, of the Occupational Safety and Health Review Commission, was assigned to the case, but it did not come to trial for more than a year. Walmart's lawyers filed twenty pretrial motions and responses and spent, by OSHA's calculations, $2 million fighting the citation. In all, OSHA lawyers invested around five thousand hours in the case. Why was Walmart fighting a paltry fine so hard? To the extent that the citation could strengthen the Damour family's civil case, $2 million could be seen as a worthwhile gamble. Moreover, no retailer welcomed OSHA jurisdiction over how it managed its customers. Casey Chroust, an executive vice president of the Retail Industry Leaders Association, told me, "The impact of this case is potentially huge. Does it mean I have to hire an event-management staff next time I hold a doorbuster sale? Does this mean every time you have a hot product—a video game, a Harry Potter book, an iPhone—much less a Black Friday sale, you'll be liable for potential action if you don't hire crowd management?" Willis Goldsmith, a partner at the New York firm of Jones Day, who has a long history of representing employers on OSHA issues, told me that along with the problem of defining crowd surges and crushes as recognized hazards, there was the practical matter of defining a crowd. "Ten people could have caused the injuries you saw at Walmart. So is that a crowd?"

OSHA's burden was to prove that crowd surge and crowd crush

are well-known phenomena and that crowd-management techniques could have prevented them at the Green Acres Mall. To do that, it needed to find an expert who would testify against Walmart. Most experts in the field consult for private industry—event planners and promoters, venue owners and operators, and, to a lesser extent, large retailers. Even if they agreed with OSHA, testifying against the world's largest retailer wasn't likely to be good for business, and many experts wouldn't do it. But one would: Paul Wertheimer, the sixty-two-year-old self-employed owner of Crowd Management Strategies, who has been called "the marshal of the mosh pit."

One of the best-documented crowd disasters in the United States occurred before a concert by the Who, outside Riverfront Coliseum in Cincinnati, on December 3, 1979. Until then, crowd planning had largely been the purview of fire-safety engineers, who focused on how to get people out of buildings in the event of an emergency—not into them. The concert's promoter, the Electric Factory of Philadelphia, had offered unreserved "festival seating" —people in the front of the line get to be nearest the stage (and, in most cases, no one on the floor has a seat at all, allowing the promoter to sell more tickets but giving the venue far less control over the audience). Hard-core fans began lining up in the early afternoon, and by six o'clock a crowd of eight thousand mostly young people had collected on the plaza outside the entrance on a bitterly cold night. The band began its sound check at around six-thirty and played for half an hour. People toward the back of the line, mistakenly believing that the concert was beginning, pushed forward. Some of the people in front pushed back, and shock waves began to ripple through the tightly packed mass. The Coliseum staff, thinking that the crowd was attempting to rush the doors and enter without paying, kept most of the doors shut, even after the sound check ended and the opening time had passed.

Later, in a letter sent to the task force assembled to investigate the incident, in which eleven people died, a man in the crowd described what it was like near the doors: "The pounding of the waves was endless. . . . If a wave came and you were being stood upon with your feet pinned to the ground, you would very likely lose your shoes or your balance and fall." Some people near the doors did go down. "They began to fall, unnoticed by all but those immediately surrounding them. People in the crowd 10 feet back from them didn't know it was happening. Their cries were impossible to hear above the roar of the crowd. . . . There was a pile of people forming, and all of the people around them were being crushed into the pile, for there was no resistance. If the person in front of you went down, then you would follow, for there was no one to lean against." Then the waves began to carry him toward the pile. "With this realization I began to add to the screaming, 'They're going down, they're going down!' I yelled repeatedly. . . . A wave swept me to the left and when I regained a stance I felt I was standing on someone. The helplessness and frustration of the moment sent a wave of panic through me. I screamed with all my strength that I was standing on someone. I couldn't move. I could only scream."

The media blamed the crowd. The Lexington, Kentucky, *Herald-Leader,* describing the "surging, primitive mob," quoted a security guard who said, "Those kids were animals." Mike Royko wrote a column for the Chicago *Sun-Times* entitled "Cincinnati Barbarism: A Rockwork Orange," blaming the "barbarians" who "stomped 11 persons to death [after] having numbed their brains on weeds, chemicals, and Southern Comfort." The promoter, Larry Magid, told *Rolling Stone,* "After all, *we* didn't trample anyone to death, we didn't step on anyone, and we didn't push anyone." Pete Townshend, the band's leader, said, "It's rock. It's not the Who. It's rock and roll. Everybody—all of us—we're all bloody responsible." In the end, no one was held accountable for the deaths.

At the time, Paul Wertheimer was a twenty-nine-year-old public information officer for the city of Cincinnati. He became chief of staff of the task force that Mayor Kenneth Blackwell appointed to investigate the incident. Wertheimer and some of his staff members spent months traveling around the country, talking to venue operators and promoters and public safety officials. Among the task force's recommendations were a ban on festival seating for large indoor events and a requirement that organizers file a "crowd management" plan, similar to a fire-safety plan, but focusing on ingress as well as egress. The report pointed out that doors and turnstiles in buildings of public assembly were tested only for normal conditions and failed to take crowded conditions into account. It also called for national standards to better protect crowds. But national standards weren't created and festival seating wasn't universally banned. Injuries and fatalities at concerts continued.

As Wertheimer worked at various jobs in event management and public relations, "the Who tragedy kept following me around," he recalled. "Every now and then, another incident would happen at a concert, someone would get killed, and the reaction was always the same. The industry would say, 'How could we have predicted this? This has never happened before!' And of course I would say, 'That's not true—it did happen, and here's a report about it!' But the industry chose to ignore that. And I thought, Somebody has to step up and do something, because there are ways to prevent these people from dying. And I guess that guy is going to be me. I am going to be the ghost of that Who concert. Those eleven people died so that these lessons could be learned, and I'm going to see they aren't forgotten."

Wertheimer began carefully documenting crowd-related incidents in the United States and around the world, making the information available to the public. He ventured into potentially dangerous crowds wherever he could find them and noted what he saw. In the early nineties, with the popularity of grunge music, mosh pits became common at rock concerts: fans in the front would hurl themselves at one another, and the force would carry them into other fans. Mosh pits are good places to study crowd dynamics, because they reproduce in miniature the shock waves of large-scale crowd disasters. Wertheimer, in his early forties, became a familiar figure at grunge and heavy-metal shows: "the old man in the pit," in the words of one young fan. "I learned how to stand in the center spot," he told me proudly, "right in front of the lead singer, three yards from the stage, and to go with the surge, and I developed my ways of getting out of tight spots, which I published in my mosh-survival guide. I worked on my peripheral vision, and learned to recognize when people are in trouble, and to understand what draws them to moshing, and how the band relates to it, and what security does in certain situations—all that stuff." He established a web site, Crowd Safe, where he published his reports on crowds, which eventually numbered in the thousands.

As predicted, none of this helped Wertheimer's career as a crowd-management consultant; his pugnacious personality didn't help, either. "The industry didn't want anything to do with me," Wertheimer told me. In Chicago, where Wertheimer was born, on the South Side, he ran afoul of a concert promoter, Jam Productions, for helping to publicize safety issues at rock concerts. (Wertheimer brought a local news reporter with a concealed camera into the mosh pit at a show put on by Jam and pointed out the unsafe conditions. Jam contends that the footage was misleading.) Jam posted Wertheimer's photograph around Soldier Field, and during a Pearl Jam concert he was picked up in a mosh pit by security for apparently shoving a young fan. "Obviously, if I wanted to develop a consulting business, this wasn't the way to do it," he told me.

After the deaths of the nine festival-goers during the 2000 Pearl Jam set in Roskilde, Wertheimer was interviewed by a committee set up by the Danish government, and recommendations he made became a part of the committee's official report, *Rock Festival Safety*. He was delighted when OSHA asked him to testify in the Walmart case. "This is the most important thing I've ever been involved with," he said. "For the first time, you've got someone powerful—the US federal government—alleging that this death was preventable, if the crowd had been handled the right way." Was he anxious about the trial? "I know you can pay a price if you take on a large corporation like Walmart. You have to be willing to suffer the consequences. I don't have kids to support, or a family; this is the role I take. I'm the only one who would do this. And, hey, I learned to fight on the South Side."

During the years that Wertheimer was recording his experiences at rock concerts, researchers in academia were trying to figure out models for crowd behavior. In the early nineties, Dirk Helbing, a graduate student in physics at the University of Göttingen, Germany, was looking for a suitable topic for a diploma thesis, when he was inspired by footprints left in the snow after a large event. He saw a pattern in the tracks that suggested the flow of streams, and he came up with a model based on fluid dynamics to simulate crowd movement. By comparing computer-driven simulations with empirical observations of crowd movement, Helbing and his colleagues were able to identify several patterns of collective behavior that emerge from the interactions of individuals in the crowd. These include lanes of uniform walking directions, oscillations of the pedestrian flow at bottlenecks, and "stripes" of intersecting flows. "Such self-organized patterns of motion demonstrate that efficient, 'intelligent' collective dynamics can be based on simple, local interactions," Helbing wrote in a 2010 paper, "Pedestrian, Crowd, and Evacuation Dynamics," published in the *Encyclopedia of Complexity and Systems Science.*

But Helbing also observed that at certain critical densities, such as occur in a crowd crush, all forms of collective behavior vanish. Shock waves are the result not of collective behavior but of the failure of it. Individuals at the back of a crowd, unable to tell what is happening up ahead, push forward, not realizing that they are injuring the people in the front. Unlike ants and fish and birds, humans haven't evolved the capability to transmit information about the physical dynamics of the crowd across the entire swarm. Ants, for example, are able to communicate within a swarm using pheromones. Iain Couzin, a behavioral biologist at Princeton University, told me, "With ants, as with human crowds, you see emergent behavior. By using a simple set of local interactions, ants form complex patterns. The difference is that we are selfish individuals, whereas ants are profoundly social creatures. We want to reduce our travel time, even when it is at the expense of others, whereas ants work for the whole colony. In this respect, we are at our most primitive in crowds. We have never evolved a collective intelligence to function in large crowds—we have no way of getting beyond the purely local rules of interaction, as ants can."

So is there no possibility that a crowd of bodies can be "smart," in the sense that a crowd of minds can be? Couzin pointed to the role that "leaders" play in the sudden movements of schools of fish or in migratory herds of animals: only a few of the animals possess the necessary information about where to go, but the others spontaneously follow them. In 2005 he helped design an experiment at Leeds University led by Jens Krause, in which two hundred people were told to walk randomly around a large hall, while a few people were given specific instructions about what route to take. The researchers found that the "naive" group followed the informed "leaders," even though they had no idea, in most cases, that they were following leaders at all. "Leadership does not require verbal communication," Couzin told me. Studies of disaster evacuations, including the 2001 World Trade Center bombing, have shown that people who follow well-informed leaders might stand a better chance of escape than people who delay or seek their own way out, but in a crowd crush that isn't going to help much. The leaders will be hemmed in too.

The Walmart trial took place during six very hot days in July 2010, in a courtroom in the Jacob Javits Federal Building in lower Manhattan. Four Walmart employees, who had been at the entrance of the vestibule with Jdimytai Damour, testified. Justin Rice, who had been promoted to department manager before Black Friday 2008 and who was still working at the store, said that the doors had broken on Blitz Day in 2007, and he had been nicked by broken glass. (Another employee said that the doors came off the hinges in 2005 and 2006 as well.) All the men said that they had never had any training in crowd management before being placed in the vestibule on November 28, 2008, except for "slip, trip, and fall" guidelines—if a customer slips, you help him up—and the "ten-foot rule": if a customer gets within ten feet you are supposed to greet her with "Welcome to Walmart."

One particularly damning bit of evidence was a video that students from the New York Institute of Technology had chanced to make of a management meeting two days before the Blitz Day event. Rice can be heard raising the matter of the 2007 melee with Steve Sooknanan, the Walmart manager, and saying that people had to be kept away from the doors this year. He says, "Last year was crazy, a lot of people fell, little babies out there and it was cold, I just don't want that this year." Sooknanan tells him that this year "we're going to do it a little differently." He explains that he had arranged for construction barriers to be placed farther from the entrance and to have additional staff at the door.

Jason Schwartz, the lead trial attorney for Walmart, wasted no time in attacking Paul Wertheimer's qualifications as a crowd expert—"the dubiously monikered 'marshal of the mosh pit'":

J.S.: What do you do when you're in a crowd, Mr. Wertheimer, in order to enhance your expertise?

P.W.: I observe the crowd, the crowd dynamics, the crowd behavior, and people in the crowd and talk to people in the crowd to see how they're feeling, see what's going on.

J.S.: If I did that, would I have the same level of experience in crowds as you do?

P.W.: No.

J.S.: Why not?
P.W.: You're not an expert in the area of crowd management.
J.S.: I see. . . . Your Honor, I would submit that this expert's qualifications are the same qualifications that everyone standing in this courtroom has.

Judge Rooney responded, "But he has more experience in crowds than I do. I don't take subways, so I have no idea what it's like to be in a crowd. Well, I could say, back in my days of college, I took the subway here in New York, and I was very claustrophobic. So I do believe that there is some assistance that, or some value that, is going to be elicited from this case."

Wertheimer was allowed to continue, and during two days of testimony detailed many crowd measures that Walmart could have taken. He was particularly effective in showing why the construction barriers wouldn't control the crowd: they were too low to keep people out, and they were flared at the bottom, so that people who got pushed up against the sides fell in.

At the end of six days, Judge Rooney had 1,200 pages of testimony to deliberate over, which she has done, at a stately pace, for the past six months. Both sides eagerly await the verdict, which is expected soon. If OSHA wins, Walmart will almost certainly appeal—all the way to the US Court of Appeals, if necessary. Still, a decision for OSHA will have enormous symbolic value, because it would be a victory for the crowd.

In the past thirty years, safety officials and designers have learned a lot about crowd management. After the Hillsborough disaster, Britain banned standing terraces in its top two soccer divisions and introduced "all-seater" stadiums. Some people argued that this changed the atmosphere of the games profoundly, but it also made them safer. An international team of experts, including Keith Still, a professor of crowd dynamics, made recommendations for the redesign of the Jamarat Bridge in Mecca and for directing the movement and flow of people. The structure has been altered to provide pilgrims with multiple entrance and exit points, to ease congestion. In Times Square on New Year's Eve, the police use lightweight metal container pens so that people revel inside a series of small enclaves rather than as one big mass. Crowd managers use elevated viewing platforms to see over the crowd and, if necessary, to communicate with people in the back. Paul Wertheimer has written a booklet, *You and the Festival Crowd,* which has been widely distributed. (Among his recommendations: Keep your elbows akimbo to protect your chest and give yourself enough breathing room. Don't fight against the flow of the crowd if you're trying to get out of it; rather, go with it, and during lulls try to work your way diagonally through the crowd to the perimeter. If you feel faint, grab on to someone, and if you do fall, try to protect your head.)

And yet almost anywhere, you can be trapped in a crowd: on a subway platform, at the lighting of the Christmas tree in Rockefeller Center, on the ramps leading down from the upper tiers at Yankee Stadium, in the Halloween parade in Greenwich Village. One reason last summer's Love Parade disaster in Germany was so shocking is that it occurred in a country known for efficient crowd management, and yet the early evidence suggests that the organizers and the police made a series of elementary mistakes, including underestimating the number of attendees, using the railway tunnel as both the main entrance to and the main exit from the event, and blocking the flow of concert-goers at pinch points, which allowed the crowd force to build. A full-scale investigation is under way.

A light rain was falling over the parking lot at the Green Acres Mall when I pulled in at three in the morning on Black Friday, 2010. The longest line was at Best Buy—it stretched the length of the building and halfway down the other side. The people in front had been waiting for twenty-eight hours. "Wii Bundles," one man said, when I asked why, as though the answer was obvious. Target also had a long line outside, and there were smaller lines outside Kohl's and Macy's. But outside Walmart there was no line at all.

After Black Friday, 2008, Walmart dropped the term Blitz Day and rebranded its post–Thanksgiving Day sale the Event. In keeping with the terms of its agreement with the Nassau County DA's office, the company employed a crowd-management plan at all its New York stores. In Valley Stream, there were more staff, security, and crowd managers outside the store than there were customers. I snaked through the barricades—metal, chest high, with open bottoms—that had been arranged in a series of tight S curves, passing two viewing platforms, with a man on each holding a bullhorn welcoming me. I entered the vestibule where Damour died, remembering the images of chaos I had seen in the videos of that night. Perhaps the most horrifying aspect of those videos is the sound inside the vestibule: cries of pain, fear, terror, mayhem. But now it was eerily quiet.

This year, like last, the waiting took place inside the store, which remained open all night. Beginning at midnight, the store began distributing tickets for the steeply discounted electronic items, and by three-fifteen they had all been given out. People arriving when I did weren't happy. "You said the sale starts at five. That's false advertising," one irate customer said to a manager. "It's not me, it's them," the manager said, gesturing toward the ceiling. People were lining up anyway for the ordinary sale-priced items, but there was no joy of the hunt in the line. It was just a line.

DAVID KIRBY

Ill Wind

FROM *Discover*

"THERE IS NO PLACE called away." It is a statement worthy of Gertrude Stein, but the University of Washington atmospheric chemist Dan Jaffe says it with conviction: none of the contamination we pump into the air just disappears. It might get diluted, blended, or chemically transformed, but it has to go somewhere. And when it comes to pollutants produced by the booming economies of East Asia, that somewhere often means right here, the mainland of the United States.

Jaffe and a new breed of global air detectives are delivering a sobering message to policymakers everywhere: carbon dioxide, the predominant driver of global warming, is not the only industrial byproduct whose effects can be felt around the world. Prevailing winds across the Pacific are pushing thousands of tons of other contaminants—including mercury, sulfates, ozone, black carbon, and desert dust—over the ocean each year. Some of this atmospheric junk settles into the cold waters of the North Pacific, but much of it eventually merges with the global air-pollution pool that circumnavigates the planet.

These contaminants are implicated in a long list of health problems, including neurodegenerative disease, cancer, emphysema, and perhaps even pandemics like avian flu. And when wind and weather conditions are right, they reach North America within days. Dust, ozone, and carbon can accumulate in valleys and basins, and mercury can be pulled to Earth through atmospheric sinks that deposit it across large swaths of land.

Pollution and production have gone hand in hand at least since the industrial revolution, and it is not unusual for a developing nation to value economic growth over environmental regulation. "Pollute first, clean up later" can be the general attitude, says Jennifer Turner, director of the China Environment Forum at the Woodrow Wilson International Center for Scholars. The intensity of the current change is truly new, however.

China in particular stands out because of its sudden role as the world's factory, its enormous population, and the mass migration of that population to urban centers; 350 million people, equivalent to the entire US population, will be moving to its cities over the next ten years. China now emits more mercury than the United States, India, and Europe combined. "What's different about China is the scale and speed of pollution and environmental degradation," Turner says. "It's like nothing the world has ever seen."

Development there is racing far ahead of environmental regulation. "Standards in the United States have gotten tighter because we've learned that ever-lower levels of air pollution affect health, especially in babies and the elderly," Jaffe says. As pollutants coming from Asia increase, though, it becomes harder to meet the stricter standards that our new laws impose.

The incoming pollution has sparked a fractious international debate. Officials in the United States and Europe have embraced the warnings of the soft-spoken Jaffe, who, with flecks of red and gray in his trim beard, looks every bit the part of a sober environmental watchdog. In China, where economic expansion has run to 8 to 14 percent a year since 2001, the same facts are seen through a different lens.

China's smog-filled cities are ringed with heavy industry, metal smelters, and coal-fired power plants, all crucial to that fast-growing economy even as they spew tons of carbon, metals, gases, and soot into the air. China's highways are crawling with the newly acquired cars of a burgeoning middle class. Still, "it's unfair to put all the blame on China or Asia," says Xinbin Feng of the Institute of Geochemistry at the Chinese Academy of Sciences, a government-associated research facility. All regions of the world contribute pollutants, he notes. And much of the emissions are generated in making products consumed by the West.

Our economic link with China makes all the headlines, but Jaffe's work shows that we are environmentally bound to the world's fastest-rising nation as well.

Dan Jaffe has been worrying about air pollution since childhood. Growing up near Boston, he liked to fish in local wetlands, where he first learned about acid rain. "I had a great science teacher, and we did a project in the Blue Hills area. We found that the acidity of the lake was rising," he recalls. The fledgling environmental investigator began chatting with fishermen around New England. "All these old-timers kept telling me the lakes had been full of fish that were now gone. That mobilized me to think about when we burn fossil fuels or dump garbage, there is no way it just goes somewhere else."

By 1997 Jaffe was living in Seattle, and his interest had taken a slant: Could pollution reaching his city be blowing in from somewhere else? "We had a hunch that pollutants could be carried across the ocean, and we had satellite imagery to show that," Jaffe says. "And we noticed our upstream neighbors in Asia were developing very rapidly. I asked the question: Could we see those pollutants coming over to the United States?"

Jaffe's colleagues considered it improbable that a concentration of pollutants high enough to significantly impact American air quality could travel thousands of miles across the Pacific Ocean; they expected he would find just insignificant traces. Despite their skepticism, Jaffe set out to find the proof. First he gathered the necessary equipment. Devices to measure carbon monoxide, aerosols, sulfur dioxide, and hydrocarbons could all be bought off the shelf. He loaded the equipment into some university trucks and set out for the school's weather observatory at Cheeka Peak. The little mountain was an arduous five-hour drive northwest of Seattle, but it was also known for the cleanest air in the Northern Hemisphere. He reckoned that if he tested this reputedly pristine air when a westerly wind was blowing in from the Pacific, the Asian pollutants might show up.

Jaffe's monitors quickly captured evidence of carbon monoxide, nitrogen oxides, ozone, hydrocarbons, radon, and particulates. Since air from North America could not have contaminated Cheeka Peak with winds blowing from the west, the next step was identifying the true source of the pollutants. Jaffe found his answer in atmospheric circulation models, created with the help of data from Earth-imaging satellites, which allowed him to trace the pollutants' path backward in time. A paper he published two years later summarized his conclusions succinctly. The pollutants "were all statistically elevated . . . when the trajectory originated over Asia."

Officials at the US Environmental Protection Agency took note, and by 1999 they were calling Jaffe to talk. They were not calling about aerosols or hydrocarbons, however, as concerning as those pollutants might be. Instead, they were interested in a pollutant that Jaffe had not looked for in his air samples: mercury.

Mercury is a common heavy metal, ubiquitous in solid material on Earth's surface. While it is trapped it is of little consequence to human health. But whenever metal is smelted or coal is burned, some mercury is released. It gets into the food chain and diffuses deep into the ocean. It eventually finds its way into fish, rice, vegetables, and fruit.

When inorganic mercury (whether from industry or nature) gets into wet soil or a waterway, sulfate-reducing bacteria begin incorporating it into an organic and far more absorbable compound called methylmercury. As microorganisms consume the methylmercury, the metal accumulates and migrates up the food chain; that is why the largest predator fish (sharks and swordfish, for example) typically have the highest concentrations. Nine-tenths of the mercury found in Americans' blood is the methyl form, and most comes from fish, especially Pacific fish. About 40 percent of all mercury exposure in the United States comes from Pacific tuna that has been touched by pollution.

In pregnant women, methylmercury can cross the placenta and negatively affect fetal brain development. Other pollutants that the fetus is exposed to can also cause toxic effects, "potentially leading to neurological, immunological, and other disorders," says the Harvard epidemiologist Philippe Grandjean, a leading authority on the risks associated with chemical exposure during early development. Prenatal exposure to mercury and other pollutants can lead to lower IQ in children—even at today's lower levels, achieved in the United States after lead paint and leaded gasoline were banned.

Among adults, the University of California, Los Angeles, neuroscience researcher Dan Laks has identified an alarming rise in mercury exposure. He analyzed data collected by the Centers for Disease Control and Prevention on 6,000 American women and found that concentrations of mercury in the human population had increased over time. Especially notable was that Laks detected inorganic mercury (the kind that doesn't come from seafood) in the blood of 30 percent of the women tested in 2005–2006, up from just 2 percent of women tested six years earlier. "Mercury's neurotoxicity is irrefutable, and there is strong evidence for an association with Alzheimer's and Parkinson's disease and amyotrophic lateral sclerosis," Laks adds.

Circumstantial evidence strongly pointed to China as the primary origin of the mercury; the industrial processes that produce the kinds of pollutants Jaffe was seeing on Cheeka Peak should release mercury as well. Still, he could not prove it from his data. To confirm the China connection and to understand the exact sources of the pollution, researchers had to get snapshots of what was happening inside that country.

One of the first scientists with feet on the ground in China was David Streets, a senior energy and environmental policy scientist at Argonne National Laboratory in Illinois. In the 1980s he was at the forefront of the study of acid rain, and in the 1990s he turned his attention to carbon dioxide and global warming as part of the Intergovernmental Panel on Climate Change. Streets began focusing on emissions from China about fifteen years ago and has since become such a noted expert that he helped the Chinese government clean up the smoke-clogged skies over Beijing before the Olympics in 2008.

In 2004, spurred by increased attention to mercury in the atmosphere, Streets decided to create an inventory of China's mercury emissions. It was a formidable undertaking. Nobody had ever come up with a precise estimate, and the Chinese government was not exactly known for its transparency.

Nevertheless, Streets considered the endeavor important because China is full of the two biggest contributors to human-generated mercury, metal smelting and coal combustion. Smelting facilities heat metal ores to eliminate contaminants and extract the desired metal, such as zinc, lead, copper, or gold. Unfortunately, one of the consistent contaminants is mercury, and the heating process allows it to escape into the atmosphere in gaseous form.

Similarly, coal contains trace amounts of mercury, which is set free during combustion at power plants.

Streets began by studying reports from China's National Bureau of Statistics. China's provinces provide the central government with detailed data on industrial production: how much coal they burn, how much zinc they produce, and so on. "China is very good at producing statistical data. It's not always one hundred percent reliable, but at least it's a start," he says. Those statistics help the Chinese government monitor the economy, but for Streets they also quantified China's mercury-laden raw materials.

The numbers from the statistics bureau told Streets the total amount of mercury that might be emitted, but he also needed to know how much actually made it into the air. To obtain that information, he turned to pollution detectives—a group of professional contacts he had met at conferences, along with graduate students who spent time in his lab. Most of the time, Chinese factories turned these "spies" away. "Factory owners had nothing to gain and a lot to lose," Streets says. "They were nervous that the results would get leaked to the government."

Yet some of Streets's moles got through by guaranteeing that the data would stay anonymous. Once inside, they took samples of raw materials—zinc ore in a smelting facility, for example—and installed chemical detectors in smokestacks. After a few days of data collection, they passed the information to Streets.

The statistics Streets collected were hardly airtight. Factory foremen and provincial officials were not above providing inflated data to make themselves look more productive, and the managers who were willing to let his inspectors take measurements were often the very ones with nothing to hide. "There's still a lot of uncertainty," Streets concedes, "but we know more than we did before."

In 2005 Streets and his team reported their first tally of human-generated mercury emissions in China, for the year 1999. The scientists estimated the amount at 590 tons (the United States emitted 117 tons). Almost half resulted from the smelting of metals—especially zinc, because its ores contain a high concentration of mercury. Coal-burning power plants accounted for another 38 percent of Chinese mercury emissions, and that percentage may be going up. As recently as 2007, China was building two new power plants a week, according to John Ashton, a climate official in the United Kingdom.

Streets's team published a subsequent inventory estimating that China's mercury emissions had jumped to 767 tons in 2003. "Mercury emissions in China have grown at about 5 to 6 percent a year," he says. "It's pretty much undeniable."

Streets had shown that China was churning out mercury, but he was left with a big uncertainty: What happened to it on its journey aloft? Finding the answer fell to Hans Friedli, a chemist at the National Center for Atmospheric Research (NCAR) who had spent thirty-three years working for Dow Chemical. Friedli had found his own path into the esoteric world of pollution forensics. Back in the early 1990s, a conversation with his neighbor, an NCAR scientist, sparked an interest in wildfires, a major source of mercury emissions. By 1998 he had a full-time job tracking the toxin for NCAR.

With its copious mercury emissions (not only from industry but also from volcanoes, wildfires, and dust storms), Asia drew Friedli's interest. China would never allow him to do aerial studies in its airspace, but in 2001 he heard about research flights off the coasts of Japan, Korea, and China designed to track dust particles emanating from the mainland. Friedli convinced the research team to take him along to measure mercury concentrations in the atmosphere. Throughout April 2001, nineteen researchers, professors, and grad students took sixteen flights aboard a cavernous retired Navy C-130 plane custom fitted with nineteen instruments for measuring pollutants like carbon monoxide, sulfur, and ozone.

During each flight, Friedli sat at his station awaiting readouts from his mercury sensor: an intake valve that sucked in air and guided it over a gold cartridge within the plane. Any mercury in the air would be absorbed by the gold. Every five minutes the instrument rapidly heated the gold, releasing any trapped mercury.

Plumes of mercury-laced air near Earth's surface are mixed with other pollutants, but at 20,000 feet Friedli discovered concentrated mercury plumes soaring eastward toward North America. He concluded that those plumes must have circled the entire globe at least once, releasing more ephemeral pollutants like carbon monoxide, so the mercury stood out even more.

Eager to follow the trail of Asian mercury plumes, Friedli set his sights across the Pacific, off the West Coast of the United States. In a series of eleven research flights in 2002, he identified a plume that looked very much like the ones he'd found near China the year before. Specifically, the plume had a ratio of carbon monoxide to mercury that served as a fingerprint for gases from the same source.

What Friedli detected was just one detail of a much larger picture. Mercury plumes can wobble in latitude and altitude or park themselves in one spot for days on end. Emissions from China —and from the United States, and indeed from every industrial country—feed a network of air currents that, as equal-opportunity polluters, serve up toxic mercury around the world.

Drawing insights from research by Friedli and Streets, Jaffe looked at his data anew. If mercury was arriving from China, he should be able to detect it, yet his operation on Cheeka Peak showed no such signal. Conducting reconnaissance from a plane, he realized why. The peak, at 1,500 feet, was below the mercury plume. Seeking a higher perch, he chose Mount Bachelor, a ski resort in central Oregon at an altitude of 9,000 feet.

In late winter 2004, Jaffe and his students huddled deep in their down jackets, bracing against a bitter gale that buffeted the chairlift ferrying them and their costly equipment to the summit. Inside the mountaintop lodge they installed a small computer lab and extended tubes outside to vacuum up the air. Later that year they conducted a similar experiment in Okinawa, Japan.

Back in Washington, they plotted their analysis of mercury in the air against satellite data showing wind currents. "My hypothesis was that we would see the same chemicals, including the same ratio of mercury to carbon monoxide, from Mount Bachelor and Japan," Jaffe says. The numbers showed exactly the expected similarity. "This was a real 'aha' moment for us, because the two regions were phenomenally close."

It was the first time anyone had decisively identified Asian mercury in American air, and the quantities were stunning. The levels Jaffe measured suggested that Asia was churning out 1,400 tons a year. The results were a shock to many scientists, Jaffe says, because "they still couldn't wrap their heads around the magnitude of the pollution and how dirty China's industry was." They were only starting to understand the global nature of the mercury problem.

Over the years, Jaffe's Mount Bachelor Observatory has also monitored many other noxious pollutants wafting across the Pacific. One major category is sulfates, associated with lung and heart disease. When sulfur dioxide exits China's coal and oil smokestacks, it converts into sulfates in the air. "Sulfates are water-soluble and get removed from the atmosphere relatively quickly, creating acid rain that falls in China, Korea, and Japan," Jaffe says. Yet some of the sulfates stay aloft, finding their way here and contributing to smog along the West Coast.

Another Chinese import is black carbon, the soot produced by cars, stoves, factories, and crop burning and a major component of Chinese haze. The small diameter of the carbon particles means they can penetrate deep inside the lungs, providing absorption sites for secondary toxins that would otherwise be cleared. This compounds the danger, making black carbon an especially potent risk factor for lung disease and premature death.

The biggest pollutant coming out of Asia, at least in terms of sheer mass, could be dust from the region's swelling deserts. "It's not a new phenomenon," Jaffe says, but it has gotten worse with deforestation and desertification caused by poorly managed agriculture. About every three years, a huge dust storm over China sends enormous clouds across the Pacific. "We can visually see it," Jaffe says. "It usually hangs around for about a week. We've tried to quantify how much it contributes to the particulate loading here, and it's a little under 10 percent of the US standard on average each year. It's a significant amount."

Chinese dust has obscured vistas in US national parks, even on the East Coast. The amount of dust is widely variable and can hit rare extreme peaks. The highest level recorded was from a 2001 dust event. "It reached approximately two-thirds of the US air quality standard at several sites along the West Coast," he reports. One study from Taiwan tracked avian flu outbreaks downwind of Asian dust storms and found that the flu virus might be transported long-distance by air spiked with the dust.

Perhaps the most counterintuitive traveling contaminant is ozone, commonly associated with ground-level pollution in cities. Volatile organic compounds, carbon monoxide, and nitrogen oxides from Asian cars and industry mix in the atmosphere as they cross the Pacific Ocean and convert in sunlight into ozone, a main ingredient in smog, Jaffe explains. When air with high ozone concentrations touches down in North America, it can pose the classic dangers of urban smog: heart disease, lung disease, and death.

Jaffe recently coauthored a paper on Asian ozone coming to

America. It found that ozone levels above western North America creep upward every spring. "When air was coming from Asia, the trend was strongest. That was the nail in the coffin," Jaffe says. "The increase was estimated at 0.5 part per billion [ppb] per year. But that's huge. In ten years that's another 5 ppb. Let's say the EPA orders a 5-ppb reduction and we achieve that, and yet, because of the growing global pool, in ten years that gets wiped out. We'll have to keep reducing our emissions just to stay even."

The underlying message of Jaffe's detective work should not be all that surprising: all of the world's atmosphere is interconnected. People have accepted this notion when it comes to carbon dioxide or the chemicals that eat away at the ozone layer, but Jaffe is finding that they are still coming to terms with the reality that it applies to industrial pollutants in general.

The fact is, those pollutants are everybody's responsibility, not just China's. The EPA has estimated that just one-quarter of US mercury emissions from coal-burning power plants are deposited within the contiguous United States. The remainder enters the global cycle. Conversely, current estimates are that less than half of all mercury deposition within the United States comes from American sources.

Then again, the United States has spent considerable effort over the past half-century trying to clean up its act. China is still much more focused on production. To fuel its boom, China has become a pioneer in wind power but has also begun buying up huge inventories of coal from markets around the world. Streets recently estimated that China's use of coal for electricity generation will rise nearly 40 percent over the next decade, from 1.29 billion tons last year to 1.77 billion tons in 2020. That is a lot more pollution to come.

"It's a classic example of a tragedy of the commons," Jaffe says, referring to a dilemma in which individuals act in their own self-interest and deplete a shared resource. "If twenty people are fishing in the same pond, with no fishing limit, then you catch as many as you can because it will be empty in weeks. Nobody has an incentive to conserve, and the same goes for pollution."

The discovery of the global mercury cycle underscores the need for an international treaty to address such pollutants. Under the auspices of the United Nations, negotiations have at least begun.

Jaffe, Streets, and China's Xinbin Feng are now consultants to the UN Environment Programme's Global Partnership on Mercury Atmospheric Transport and Fate Research, which helped contribute data that led to a proposed UN mercury treaty in 2009.

When it comes to some pollutants, China has taken important steps. For instance, recent policies encourage desulfurization and other filtering technology in power plants. But convincing developing nations to move aggressively on mercury may be at least as tough as mobilizing them against carbon emissions. "This is not considered a pollutant that urgently needs to be controlled on the national level," Feng says. "It's not fair that you emitted so much mercury and other pollutants when you had the chance to industrialize. You had two hundred years, and now you want to stop other countries from developing too."

"We need to be concerned," Jaffe counters in his low-key way. "There is no Planet B. We all live downwind."

ROBERT KUNZIG

The City Solution

FROM *National Geographic*

AT THE TIME of Jack the Ripper, a hard time for London, there lived in that city a mild-mannered stenographer named Ebenezer Howard. He's worth mentioning because he had a large and lingering impact on how we think about cities. Howard was bald, with a bushy, mouth-cloaking mustache, wire-rim spectacles, and the distracted air of a seeker. His job transcribing speeches did not fulfill him. He dabbled in spiritualism; mastered Esperanto, the recently invented language; invented a shorthand typewriter himself. And dreamed about real estate. What his family needed, he wrote to his wife in 1885, was a house with "a really nice garden with perhaps a lawn tennis ground." A few years later, after siring four children in six years in a cramped rental house, Howard emerged from a prolonged depression with a scheme for emptying out London.

London in the 1880s, you see, was booming, but it was also bursting with people far more desperate than Howard. The slums where the Ripper trolled for victims were beyond appalling. "Every room in these rotten and reeking tenements houses a family, often two," wrote Andrew Mearns, a crusading minister. "In one cellar a sanitary inspector reports finding a father, mother, three children, and four pigs! . . . Elsewhere is a poor widow, her three children, and a child who had been dead thirteen days." The Victorians called such slums rookeries, or colonies of breeding animals. The chairman of the London County Council described his city as "a tumour, an elephantiasis sucking into its gorged system half the life and the blood and the bone of the rural districts."

Urban planning in the twentieth century sprang from that horrified perception of nineteenth-century cities. Oddly, it began with Ebenezer Howard. In a slim book, self-published in 1898, the man who spent his days transcribing the ideas of others articulated his own vision for how humanity ought to live—a vision so compelling that half a century later Lewis Mumford, the great American architecture critic, said it had "laid the foundation for a new cycle in urban civilization."

The tide of urbanization must be stopped, Howard argued, by drawing people away from the cancerous metropolises into new, self-contained "garden cities." The residents of these happy little islands would feel the "joyous union" of town and country. They'd live in nice houses and gardens at the center, walk to work in factories at the rim, and be fed by farms in an outer greenbelt—which would also stop the town from expanding into the country. When one town filled to its greenbelt—32,000 people was the right number, Howard thought—it would be time to build the next one. In 1907, welcoming 500 Esperantists to Letchworth, the first garden city, Howard boldly predicted (in Esperanto) that both the new language and his new utopias would soon spread around the world.

He was right about the human desire for more living space but wrong about the future of cities: it's the tide of urbanization that has spread around the world. In the developed countries and Latin America it has nearly crested; more than 70 percent of people there live in urban areas. In much of Asia and Africa, people are still surging into cities, in numbers swollen by the population boom. Most urbanites live in cities of less than half a million, but big cities have gotten bigger and more common. In the nineteenth century, London was the only city of more than 5 million; now there are fifty-four, most of them in Asia.

And here's one more change since then: urbanization is now good news. Expert opinion has shifted profoundly in the past decade or two. Though slums as appalling as Victorian London's are now widespread, and the Victorian fear of cities lives on, cancer no longer seems the right metaphor. On the contrary: with Earth's population headed toward 9 or 10 billion, dense cities are looking more like a cure—the best hope for lifting people out of poverty without wrecking the planet.

*

One evening last March, the Harvard economist Edward Glaeser appeared at the London School of Economics to promote this point of view, along with his new book, *Triumph of the City.* Glaeser, who grew up in New York City and talks extremely fast, came heavily armed with anecdotes and data. "There's no such thing as a poor urbanized country; there's no such thing as a rich rural country," he said. A cloud of country names, each plotted by GDP (gross domestic product) and urbanization rate, flashed on the screen behind him.

Mahatma Gandhi was wrong, Glaeser declared—India's future is not in its villages, it's in Bangalore. Images of Dharavi, Mumbai's large slum, and of Rio de Janeiro's favelas flashed by; to Glaeser, they were examples of urban vitality, not blight. Poor people flock to cities because that's where the money is, he said, and cities produce more because "the absence of space between people" reduces the cost of transporting goods, people, and ideas. Historically, cities were built on rivers or natural harbors to ease the flow of goods. But these days, since shipping costs have declined and service industries have risen, what counts most is the flow of ideas.

The quintessence of the vibrant city for Glaeser is Wall Street, especially the trading floor, where millionaires forsake large offices to work in an open-plan bath of information. "They value knowledge over space—that's what the modern city is all about," he said. Successful cities "increase the returns to being smart" by enabling people to learn from one another. In cities with higher average education, even the uneducated earn higher wages; that's evidence of "human capital spillover."

Spillover works best face-to-face. No technology yet invented—not the telephone, the Internet, or videoconferencing—delivers the fertile chance encounters that cities have delivered ever since the Roman Forum was new. Nor do they deliver the nonverbal, contextual cues that help us convey complex ideas—to see from the glassy eyes of our listeners, for instance, that we're talking too fast.

It's easy to see why economists would embrace cities, warts and all, as engines of prosperity. It has taken a bit longer for environmentalists, for whom Henry David Thoreau's cabin in the woods has been a lodestar. By increasing income, cities increase consumption and pollution too. If what you value most is nature, cities look like concentrated piles of damage—until you consider the alternative, which is spreading the damage. From an ecological standpoint, says Stewart Brand, founder of the *Whole Earth Catalog* and now a champion of urbanization, a back-to-the-land ethic would be disastrous. (Thoreau, Glaeser points out gleefully, once accidentally burned down three hundred acres of forest.) Cities allow half of humanity to live on around 4 percent of the arable land, leaving more space for open country.

Per capita, city dwellers tread more lightly in other ways as well, as David Owen explains in *Green Metropolis*. Their roads, sewers, and power lines are shorter and so use fewer resources. Their apartments take less energy to heat, cool, and light than do houses. Most important, people in dense cities drive less. Their destinations are close enough to walk to, and enough people are going to the same places to make public transit practical. In cities like New York, per capita energy use and carbon emissions are much lower than the national average.

Cities in developing countries are even denser and use far fewer resources. But that's mostly because poor people don't consume a lot. Dharavi may be a "model of low emissions," says David Satter-

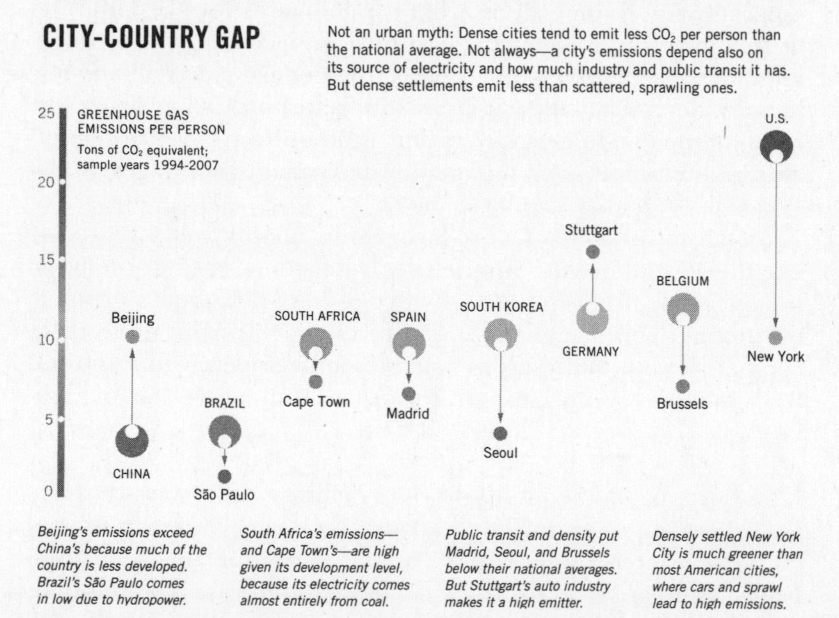

CITY-COUNTRY GAP

Not an urban myth: Dense cities tend to emit less CO_2 per person than the national average. Not always—a city's emissions depend also on its source of electricity and how much industry and public transit it has. But dense settlements emit less than scattered, sprawling ones.

Beijing's emissions exceed China's because much of the country is less developed. Brazil's São Paulo comes in low due to hydropower.

South Africa's emissions—and Cape Town's—are high given its development level, because its electricity comes almost entirely from coal.

Public transit and density put Madrid, Seoul, and Brussels below their national averages. But Stuttgart's auto industry makes it a high emitter.

Densely settled New York City is much greener than most American cities, where cars and sprawl lead to high emissions.

John Tomano, NGM Staff. Source: Dan Hoornweg, World Bank

thwaite of London's International Institute for Environment and Development, but its residents lack safe water, toilets, and garbage collection. So do perhaps a billion other city dwellers in developing countries. And it is such cities, the United Nations projects, that will absorb most of the world's population increase between now and 2050—more than 2 billion people. How their governments respond will affect us all.

Many are responding the way Britain did to the growth of London in the nineteenth century: by trying to make it stop. A UN survey reports that 72 percent of developing countries have adopted policies designed to stem the tide of migration to their cities. But it's a mistake to see urbanization itself as evil rather than as an inevitable part of development, says Satterthwaite, who advises governments and associations of slum dwellers around the world. "I don't get scared by rapid growth," he says. "I meet African mayors who tell me, 'There are too many people moving here!' I tell them, 'No, the problem is your inability to govern them.'"

There is no single model for how to manage rapid urbanization, but there are hopeful examples. One is Seoul, the capital of South Korea.

Between 1960 and 2000, Seoul's population zoomed from fewer than 3 million to 10 million, and South Korea went from being one of the world's poorest countries, with a per capita GDP of less than $100, to being richer than some countries in Europe. The speed of the transformation shows. Driving into Seoul on the highway along the Han River, you pass a distressingly homogeneous sea of concrete apartment blocks, each emblazoned with a large number to distinguish it from its clones. Not so long ago, though, many Koreans lived in shanties. The apartment blocks may be uninspiring on the outside, the urban planner Yeong-Hee Jang told me, but life inside "is so warm and convenient." She repeated the word "warm" three times.

Every city is a unique mix of the planned and the unplanned, of features that were intentionally designed by government and others that emerged organically, over time, from choices made by the residents. Seoul was planned from the start. The monks who chose the site in 1394 for King Taejo, founder of the Choson dynasty, followed the ancient principles of feng shui. They placed the king's palace at an auspicious spot, with the Han River in front and a large mountain in back to shield it from the north wind. For five centuries the city stayed mostly inside a ten-mile-long wall that Taejo's men had built in six months. It was a cloistered, scholarly town of a few hundred thousand. Then the twentieth century cleaned its slate.

World War II and then the Korean War, which ended in 1953, brought more than a million refugees to the bombed-out city. Not much of Seoul was left—but it was filled for the first time with a potent mix of people. They were burning to improve their miserable lot. In their hearts, the ancient Confucian virtues of loyalty and respect for hierarchy fused uneasily with Western longings for democracy and material goods. "The explosive energy of my generation," says Hong-Bin Kang, a former vice mayor who now runs Seoul's history museum, dates from this period. So does South Korea's population explosion, which was triggered, as elsewhere, by rapid improvements in public health and nutrition.

It's an uncomfortable fact that a dictator helped organize all that energy. When Park Chung-Hee took power in a military coup in 1961, his government funneled foreign capital into Korean companies that made things foreigners would buy—knockoff clothes and wigs at first, later steel, electronics, and cars. Central to the process, which created conglomerates like Samsung and Hyundai, were the women and men streaming into Seoul to work in its new factories and educate themselves at its universities. "You can't understand urbanization in isolation from economic development," says economist Kyung-Hwan Kim of Sogang University. The growing city enabled the economic boom, which paid for the infrastructure that helped the city absorb the country's growing population.

A lot was lost in the bulldozing, high-rising rush. If you lived in old Seoul, north of the Han River, in the 1970s and 1980s, you watched an entirely new Seoul rise from verdant paddies on the south bank, in the area called Kangnam. You watched the city's growing middle and upper classes leave sinuous alleys and traditional houses—lovely wooden *hanok,* with courtyards and gracefully curved tile roofs—for antiseptic high-rises and a grid of car-friendly boulevards. "Seoul lost its color," says Choo Chin Woo, an investigative journalist at the newsweekly *SisaIN.* "Apartment high-rises all over town—it looks stupid." Worse, the poor often got shunted aside as their makeshift neighborhoods were redeveloped with high-rises they couldn't afford.

But over the years an increasing share of the population has been able to cash in on the housing boom. Today half the people in Seoul own apartments. Koreans like to heat their homes to 77 degrees, says urban planner Yeong-Hee Jang, and in their well-equipped apartments they can afford to do that. One reason the buildings in Kangnam line up like soldiers on parade, she adds, is that everyone wants an apartment that faces south—for warmth as well as feng shui.

Seoul today is one of the densest cities in the world. It has millions of cars but also an excellent subway system. Even in the newer districts, the streets seem, to a Westerner, anything but colorless. They're vibrant with commerce and crowded with pedestrians, each of whom has a carbon footprint less than half the size of a New Yorker's. Life has gotten much better for Koreans as the country has gone from 28 percent urban in 1961 to 83 percent today. Life expectancy has increased from fifty-one years to seventy-nine—a year longer than for Americans. Korean boys now grow six inches taller than they used to.

South Korea's experience can't be easily copied, but it does prove that a poor country can urbanize successfully and incredibly fast. In the late 1990s Kyung-Hwan Kim worked for the UN in Nairobi, advising African cities on their staggering financial problems. "Every time I visited one of these cities I asked myself, What would a visiting consultant have said to Koreans in 1960?" he says. "Would he have imagined Korea as it was forty years later? The chances are close to zero."

The fear of urbanization has not been good for cities, or for their countries, or for the planet. South Korea, ironically, has never quite shaken the notion that its great capital is a tumor sucking life from the rest of the country. Right now the government is building a second capital seventy-five miles to the south; starting in 2012, it plans to move half its ministries there and to scatter other public institutions around the country, in the hope of spreading Seoul's wealth. The nation's efforts to stop Seoul's growth go back to Park Chung-Hee, the dictator who jump-started the economy. In 1971, as the city's population was skyrocketing past 5 million, Park took a page from the book of Ebenezer Howard. He surrounded the city with a wide greenbelt to halt further development, just as London had in 1947.

Both greenbelts preserved open space, but neither stopped the growth of the city; people now commute from suburbs that leapfrogged the restraints. "Greenbelts have had the effect of pushing people farther out, sometimes absurdly far," says Peter Hall, a planner and historian at University College London. Brasília, the planned capital of Brazil, was designed for 500,000 people; 2 million more now live beyond the lake and park that were supposed to block the city's expansion. When you try to stop urban growth, it seems, you just amplify sprawl.

Sprawl preoccupies urban planners today, as its antithesis, density, did a century ago. London is no longer decried as a tumor, but Atlanta has been called "a pulsating slime mold" (by James Howard Kunstler, a colorful critic of suburbia) on account of its extreme sprawl. Greenbelts aren't the cause of sprawl; most cities don't have them. Other government policies, such as subsidies for highways and home ownership, have coaxed the suburbs outward. So has that other great shaper of the destiny of cities: the choices made by individual residents. Ebenezer Howard was right about that much: a lot of people want nice houses with gardens.

Sprawl is not just a Western phenomenon. By consulting satellite images, old maps, and census data, Shlomo Angel, an urban planning professor at New York University and Princeton, has tracked how 120 cities changed in shape and population density between 1990 and 2000. Even in developing countries, most cities are spreading out faster than people pour into them; on average they're getting 2 percent less dense each year. By 2030 their built-up area could triple. What's driving the expansion? Rising incomes and cheap transportation. "When income rises, people have money to buy more space," Angel explains. With cheap transportation, they can afford to travel longer distances from home to work.

But it matters what kind of homes they live in and what transportation they use. In the twentieth century, American cities were redesigned around cars—wonderful, liberating machines that also make city air unbreathable and carry suburbs beyond the horizon. Car-centered sprawl gobbles farmland, energy, and other resources. These days, planners in the United States want to repopulate downtowns and densify suburbs, by building walkable town centers, for instance, in the parking lots of failed malls. Urban flight, which seemed a good idea a century ago, now seems in the West like a historic wrong turn. Meanwhile, in China and India, where people are still flooding into cities, car sales are booming. "It would be a lot better for the planet," Edward Glaeser writes, if people in those countries end up "in dense cities built around the elevator, rather than in sprawling areas built around the car."

Developing cities will inevitably expand, says Angel. Somewhere between the anarchy that prevails in many today and the utopianism that has often characterized urban planning lies a modest kind of planning that could make a big difference. It requires looking decades ahead, Angel says, and reserving land, before the city grows over it, for parks and a dense grid of public-transit corridors. It starts with looking at growing cities in a positive way—not as diseases but as concentrations of human energy to be organized and tapped.

With its quiet commercial streets and Arts and Crafts houses, Letchworth, England, today feels a bit like the garden city that time forgot. Ebenezer Howard's ideal of a self-sustaining community never happened. The farmers in Letchworth's greenbelt sell their sugar beets and wheat to a large cereal company. The town's residents work mostly in London or Cambridge. John Lewis, who runs the foundation that Howard started, which still owns much of the town's land, worries that Letchworth is "in danger of becoming a dormitory." Still, it has a key aspect of what many planners today think of as sustainability: it wasn't designed around cars. Howard ignored the new invention. From anywhere in Letchworth you can walk to the center of town to shop or take the train to London. The truth is, Letchworth looks like a very nice place to live; it's just not for everyone. No place is.

Thirty-five miles to the south, London remains unsupplanted. Eight million people live there now. All attempts to impose sense on its maze of streets have failed, as anyone who has crossed the city in a taxi can attest. "London wasn't planned at all!" Peter Hall exclaimed one afternoon as we stepped into the street in front of the British Academy. But the city did two sensible things as it ballooned outward in the nineteenth and twentieth centuries, Hall said. It preserved large, semiwild parks like Hampstead Heath, where citizens can commune with nature. Most important, it expanded along railway and subway lines. "Get the transportation right," said Hall. "Then let things happen."

With that he disappeared into the Underground for his ride home, leaving me on the crowded sidewalk with a great gift: a few hours to kill in London. Even Ebenezer Howard would have understood the feeling, at least as a young man. When he returned after a few years in the United States—he'd flopped as a homesteading farmer in Nebraska—he was jazzed by his native city. Just riding an omnibus, he later wrote, gave him a pleasantly visceral jolt: "A strange ecstatic feeling at such times often possessed me . . . The crowded streets—the signs of wealth and prosperity—the bustle—the very confusion and disorder appealed to me, and I was filled with delight."

PART SIX

Technology

MICHAEL SPECTER

Test-Tube Burgers

FROM *The New Yorker*

WILLEM VAN EELEN was born in 1923, the son of a doctor and a child of colonial privilege. His father had recently been dispatched to the Dutch East Indies, and van Eelen wanted for nothing. "I was a spoiled boy and gave little thought to the world around me," he said not long ago, as we sat in the study of his modest apartment, which overlooks the broad waters of the Amstel River, in Amsterdam. His youth of oblivious freedom ended abruptly on May 10, 1940—the day the Nazis invaded the Netherlands. Van Eelen was just sixteen, but like many of his contemporaries, he lied about his age, enlisted, and served in Indonesia.

The Dutch fought frantically to prevent Japan from seizing their most valuable colony, but they failed. Van Eelen was captured, and spent most of the war as a prisoner, dragged forcibly from one POW camp to the next. Now, at eighty-seven, dressed in khakis, penny loafers, and a casual gray shirt, he projects the contemplative air of a philosopher. Van Eelen is a genial man who laughs easily. But when asked about the camps, he lowered his voice and slowly closed his eyes.

"These were cruel places," he said. "We worked from morning to night building airstrips. They beat us like dogs. For food, there was almost nothing. The Japanese were harsh with us, but they treated animals even more brutally, kicking them, shooting them. By the time the Americans liberated the camp, I was so close to death that you could see my spine from the front. The soldiers would ask my name, but I didn't have enough strength to say the words."

After the war, van Eelen studied psychology at the University of Amsterdam, but he struggled with the intertwined memories of starvation and animal abuse. He began to attend scientific lectures, and during one of them, about how to preserve meat, van Eelen was seized by an idea: "I wondered, Why can't we grow meat outside of the body? Make it in a laboratory, as we make so many other things." He went on, "I like meat—I never became a vegetarian. But it is hard to justify the way animals are treated on this planet. Growing meat without inflicting pain seemed a natural solution."

"Meat" is a vague term and can be used to refer to many parts of an animal, including internal organs and skin. For the most part, the meat we eat consists of muscle tissue taken from farm animals, whether it's a sirloin steak, which is cut from the rear of a cow, or a pork chop, taken from flesh near the spine of a pig. In vitro meat, however, can be made by placing a few cells in a nutrient mixture that helps them proliferate. As the cells begin to grow together, forming muscle tissue, they are attached to a biodegradable scaffold, just as vines wrap around a trellis. There the tissue can be stretched and molded into food, which could, in theory at least, be sold, cooked, and consumed like any processed meat—hamburger, for example, or sausage.

"This became my fixation," van Eelen continued. "Everything I have done since that day I have done with this goal in mind." After university, van Eelen went to medical school, where he spoke to biologists, research scientists, and anyone else he thought could help. Most people laughed when they heard about his project—in part, perhaps, because van Eelen is more of a scientific enthusiast than a sophisticate. When he told his professors that he wanted to grow meat in a lab, most acted as if it were a prank. But one teacher took him aside. "He said if I was serious I would need to raise money for research," van Eelen recalled. He promptly quit his medical studies and went to work. With his wife (an artist, who died many years ago), he ran a series of art galleries and restaurants. The couple funneled whatever money they managed to save into his odd obsession.

Van Eelen has been chasing his goal ever since, but it took decades for the science to catch up with his imagination. That began to happen in 1981, when stem cells, which can divide almost endlessly and have the ability to develop into many types of tissue, were discovered in mice. Van Eelen recognized the potential immediately, although there was little initial interest in turning muscle cells into meat. By then he was used to rejection, and he persisted. Finally, in 1999, more than half a century after he attended the lecture that fueled his quest, he received US and international patents for the Industrial Production of Meat Using Cell Culture Methods. For the first time, serious people began to take him seriously. Pointing to the channel waters outside his window, van Eelen said, "For all those years, there was not one gram of meat made. At times, I wanted to jump right into that river."

He no longer feels that way, and for good reason: a new discipline, propelled by an unlikely combination of stem-cell biologists, tissue engineers, animal rights activists, and environmentalists, has emerged in both Europe and the United States. The movement started fitfully but intensified when, in 2001, NASA funded an experiment, led by Morris Benjaminson, that focused on producing fresh meat for space flights. Benjaminson, a biological engineer at Touro College, in New York, cut strips of flesh from live goldfish and submerged them in a nutrient bath extracted from the blood of unborn cows. Within a week, the fish pieces had grown by nearly 15 percent. While the results were not meat, they demonstrated that growing food outside the body was possible. Then in 2004, after continued lobbying from van Eelen, the Dutch government awarded 2 million euros to a consortium of universities and research facilities in Amsterdam, Utrecht, and Eindhoven. Though the grant was small, it has helped turn the Netherlands into the in-vitro-meat world's version of Silicon Valley.

Van Eelen was not the only man undaunted by indifference to the idea of lab-grown meat. Vladimir Mironov, an associate professor in the Department of Cell Biology and Anatomy at the Medical University of South Carolina, is working on several experiments, most of which focus on finding an efficient way to grow it. Mironov, a well-known tissue researcher, was brought up in Russia and studied at the Max Planck Institute with the pioneering vascular biologist Werner Risau. Then, in the early 1980s, he moved to the United States, where he became intrigued by the possibilities of making meat. "A few years ago, I tried to get a grant," Mironov told me when I visited his lab in Charleston. "I failed. I tried to get venture capital. Failed again. I tried to approach big companies for funding. Failed again. But slowly, very slowly, people are coming around."

Teams are forming at universities around the world. Some are interested primarily in animal welfare, others in regenerative medicine; still others see lab meat as a potential solution to an environmental crisis. They all share a goal, however: to grow muscle without the use of animals and to produce enough of it to be sold in grocery stores. "This is a no-brainer," Ingrid Newkirk, the cofounder and president of People for the Ethical Treatment of Animals (PETA), told me. Three years ago, the animal rights organization, which has a singular gift for public relations, offered $1 million to the first group that could create "an in-vitro chicken-meat product that has a taste and texture indistinguishable from real chicken flesh." More recently, PETA provided funding for Nicholas Genovese, a postdoctoral biological engineer, to work in Mironov's lab—a sort of PETA fellowship. Newkirk explained, "If people are unwilling to stop eating animals by the billions, then what a joy to be able to give them animal flesh that comes without the horror of the slaughterhouse, the transport truck, and the mutilations, pain, and suffering of factory farming."

Meat supplies a variety of nutrients—among them iron, zinc, and Vitamin B_{12}—that are not readily found in plants. We can survive without it; millions of vegetarians choose to do so, and billions of others have that choice imposed upon them by poverty. But for at least 2 million years, animals have provided our most consistent source of protein. For most of that time, the economic, social, and health benefits of raising and eating livestock were hard to dispute. The evolutionary biologist Richard Wrangham argues, in his book *Catching Fire: How Cooking Made Us Human,* that the development of a brain that could conceive of cooking meat—a singularly efficient way to consume protein—has defined our species more clearly than any other characteristic. Animals have always been essential to human development. Sir Albert Howard, who is often viewed as the founder of the modern organic farming movement, put it succinctly in his 1940 mission statement, *An Agricultural Testament:* "Mother earth never attempts to farm without livestock."

For many people, the idea of divorcing beef from a cow or pork from a pig will seem even more unsettling than the controversial yet utterly routine practice of modifying crops with the tools of molecular biology. The Food and Drug Administration currently has before it an application, which has already caused rancorous debate, to engineer salmon with a hormone that will force the fish to grow twice as fast as normal. Clearly, making meat without animals would be a more fundamental departure. How we grow, prepare, and eat our food is a deeply emotional issue, and lab-grown meat raises powerful questions about what most people see as the boundaries of nature and the basic definitions of life. Can something be called chicken or pork if it was born in a flask and produced in a vat? Questions like that have rarely been asked and have never been answered.

Still, the idea itself is not new. On January 17, 1912, the Nobel Prize–winning biologist Alexis Carrel placed tissue from an embryonic chicken heart in a bath of nutrients. He kept it beating in his laboratory at the Rockefeller Institute for more than twenty years, demonstrating that it was possible to keep muscle tissue alive outside the body for an extended period. Laboratory meat has also long been the subject of dystopian fantasy and literary imagination. In 1931 Winston Churchill published an essay, "Fifty Years Hence," in which he described what he saw as the inevitable future of food: "We shall escape the absurdity of growing a whole chicken in order to eat the breast or wing." He added, "Synthetic food will, of course, also be used in the future. Nor need the pleasure of the table be banished. . . . The new foods will from the outset be practically indistinguishable from the natural products." The idea has often been touched on in science fiction. In *Neuromancer,* William Gibson's 1984 novel, artificial meat—called vat-grown flesh—is sold at lower prices than the meat from living animals. In Margaret Atwood's *Oryx and Crake,* published in 2003, "ChickieNobs" are engineered to have many breasts and no brains.

Past discussions have largely been theoretical, but our patterns of meat consumption have become increasingly dangerous for both individuals and the planet. According to the United Nations Food and Agriculture Organization, the global livestock industry is responsible for nearly 20 percent of humanity's greenhouse-gas emissions. That is more than all cars, trains, ships, and planes combined. Cattle consume nearly 10 percent of the world's freshwater resources, and 80 percent of all farmland is devoted to the production of meat. By 2030 the world will likely consume 70 percent more meat than it did in 2000. The ecological implications are daunting, and so are the implications for animal welfare: billions of cows, pigs, and chickens spend their entire lives crated, boxed, or force-fed grain in repulsive conditions on factory farms. These animals are born solely to be killed, and between the two events they are treated like interchangeable parts in a machine, as if a chicken were a spark plug, and a cow a drill bit.

The consequences of eating meat, and our increasing reliance on factory farms, are almost as disturbing for human health. According to a report issued recently by the American Public Health Association, animal waste from industrial farms "often contains pathogens, including antibiotic-resistant bacteria, dust, arsenic, dioxin and other persistent organic pollutants." Seventy percent of all antibiotics and related drugs consumed in the United States are fed to hogs, poultry, and beef. In most cases, they are used solely to promote growth and not for any therapeutic reason. By eating animals, humans have exposed themselves to SARS, avian influenza, and AIDS, among many other viruses. The World Health Organization has attributed a third of the world's deaths to the twin epidemics of diabetes and cardiovascular disease, both greatly influenced by excessive consumption of animal fats.

"We have an opportunity to reverse the terribly damaging impact that eating animals has had on our lives and on this planet," Mark Post, a professor in the physiology department at Maastricht University, in the Netherlands, told me. "The goal is to take the meat from one animal and create the volume previously provided by a million animals." Post, who is a vascular biologist and a surgeon, also has a doctorate in pulmonary pharmacology. His area of expertise is angiogenesis—the growth of new blood vessels. Until recently, he had dedicated himself to creating arteries that could replace and repair those in a diseased human heart. Like many of his colleagues, he was reluctant to shift from biomedicine to the meat project. "I am a scientist, and my family always respected me for that," he said. "When I started basically spending my time trying to make the beginning of a hamburger, they would give me a pitiful look, as if to say, You have completely degraded yourself."

We met recently at the Eindhoven University of Technology, where he served on the faculty for years and remains a vice-dean. "First people ask, 'Why would anyone want to do this?'" he said.

"The initial position often seems to be a reflex: nobody will ever eat this meat. But in the end I don't think that will be true. If people visited a slaughterhouse, then visited a lab, they would realize this approach is so much healthier." He added, "I have noticed that when people are exposed to the facts, to the state of the science, and why we need to look for alternatives to what we have now, the opposition is not so intense."

Post, a trim fifty-three-year-old man in rimless glasses and a polo shirt, stressed, too, that scientific advances have been robust. "If what you want is to grow muscle cells and produce a useful source of animal protein in a lab, well, we can do that today," he said—an assertion echoed by Mironov in South Carolina and by many other scientists in the field. To grow ground meat—which accounts for half the meat sold in the United States—one needs essentially to roll sheets of two-dimensional muscle cells together and mold them into food. A steak would be much harder. That's because before scientists can manufacture meat that looks as if it came from a butcher, they will have to design the network of blood vessels and arteries required to ferry nutrients to the cells. Even then, no product with a label that said "Born in cell culture, raised in a vat" would be commercially viable until the costs fall.

Scientific advances necessarily predate the broad adoption of any technology—often by years. Post points to the first general-purpose computer, Eniac. Built during World War II and designed to calculate artillery-firing ranges, the computer cost millions of dollars and occupied a giant room in the US Army's Ballistic Research Laboratory. "Today, any cell phone or five-dollar watch has a more powerful computer," Post noted. In the late 1980s, as the Human Genome Project got under way, researchers estimated that sequencing the genome of a single individual would take fifteen years and cost $3 billion. The same work can now be done in twenty-four hours for about $1,000.

Those numbers will continue to fall as personal genomics becomes more relevant, and, as would be the case with laboratory meat, it will become more relevant if the price keeps falling. "The first hamburger will be incredibly expensive," Post said. "Somebody calculated five thousand dollars. The skills you need to grow a small amount of meat in a laboratory are not necessarily those that would permit you to churn out ground beef by the ton. To do that will require money and public interest. We don't have enough of either right now. That I do not understand, because while I am no businessman, there certainly seems to be a market out there."

Meat and poultry dominate American agriculture, with sales that exceeded $150 billion in 2009. It is unlikely that the industry would cheer on competitors who could directly challenge its profits. Yet if even a small percentage of customers switched their allegiance from animals to vats, the market would be huge. After all, the world consumes 285 million tons of meat every year—90 pounds per person. The global population is expected to rise from 7 billion to more than 9 billion by the year 2050. This increase will be accompanied by a doubling of the demand for meat and a steep climb in the greenhouse-gas emissions for which animals are responsible. Owing to higher incomes, urbanization, and growing populations—particularly in emerging economies—demand for meat is stronger than it has ever been. In countries like China and India, moving from a heavily plant-based diet to one dominated by meat has become an essential symbol of a middle-class life.

Cultured meat, if it was cheap and plentiful, could dispense with many of these liabilities by providing new sources of protein without inflicting harm on animals or posing health risks to humans. One study, completed last year by researchers at Oxford and the University of Amsterdam, reported that the production of cultured meat could consume roughly half the energy and occupy just 2 percent of the land now devoted to the world's meat industry. The greenhouse gases emitted by livestock, now so punishing, would be negligible. The possible health benefits would also be considerable. Eating meat that was engineered rather than taken from an animal might even be good for you. Instead of committing slow suicide by overdosing on saturated fat, we could begin to consume meat infused with omega-3 fatty acids—which have been demonstrated to prevent the type of heart disease caused by animal fats. "I can well envision a scenario where your doctor would prescribe hamburgers rather than prohibit them," Post said. "The science is not simple, and there are hurdles that remain. But I have no doubt we will get there."

For at least a century, Eindhoven has been a technical town—first as a base for electronics, then as a center for automobile and truck manufacturing. In the past decade, it has become the capital of the Netherlands' influential industrial-design movement. When I was there, the city was filled with men and women cycling purposefully through the streets, many in dark clothing and angular eyewear. Philips, the Dutch electronics giant, was once based in the center of town, and the company's highly respected design center is still there. As architecture, industrial design, engineering, and biology have become increasingly interrelated, Eindhoven became a natural home for the nation's premier University of Technology. In turn, the university, and particularly its department of biomedical engineering, has become the hub for research into growing meat.

Soon after I arrived, Daisy van der Schaft, a thirty-four-year-old assistant professor, took me to the lab where the meat team conducts most of its experiments. Until recently she concentrated on regenerative medicine, but in vitro meat has begun to occupy increasing amounts of her time and imagination. "On the practical level, there was some grant money," she said. "And, on the personal level, this is an opportunity to do something worthwhile. But for a scientist, it's not that big a switch."

In the past decade, the idea of taking healthy cells from our own bodies and using them to grow replacement parts has moved from a hopeful theory to an increasingly frequent reality. With organ-donor shortages as a powerful incentive, medical researchers have had success in creating whole and partial organs to repair and, in some cases, replace diseased tissues. Scientists have used stem cells to construct windpipes, skin, cartilage, and bone. Biologically engineered bladders have been placed in many patients. (Anthony Atala, the director of the Wake Forest Institute for Regenerative Medicine, described, in a talk he delivered at the TED Conference in March, how he had implanted artificially constructed bladders in people who subsequently were healthy for years. While Atala spoke in an auditorium in Long Beach, California, a three-dimensional printer was busy in the background, producing the prototype of a kidney. Instead of ink, however, the printer used layers of cells that it then fused together.) In Tokyo, scientists have developed a technique for wrapping a thin sheet of cardiomyocytes —muscle cells that the heart needs in order to beat—around the severely damaged hearts of patients. Once implanted, the sheets, beating independently, act like an extra battery. Such successes have helped spark interest in the meat project, because the skills required to fashion an organ from stem cells are similar to those needed to make minced meat or sausage in a petri dish.

Van der Schaft handed me a starched white coat and pointed to a row of incubators—delivery rooms used by researchers to grow cells and tissues of all kinds. "It's an exciting project," she said as she reached into an incubator and removed one of many small Plexiglas boxes. "A hopeful project." Each box contained six disks filled with muscle cells. The cells, gelatinous brown smears resting between identical Velcro beds filled with nutrients, were nearly impossible to see without a microscope. "This is what I have to show you right now," she said, grimacing. "They did tell you we didn't have meat as such, right?" I had been duly informed. The team learned long ago that visitors feel cheated when they realize that there will be no lunch of faux chicken or vat-ripened pork. Despite the warning, I felt cheated too.

Nearly every person I told that I was working on this piece asked the same question: What does it taste like? (And the first word most people blurted out to describe their feelings was "Yuck.") Researchers say that taste and texture—fats and salt and varying amounts of protein—can be engineered into lab-grown meat with relative ease. For the moment, taste remains a secondary issue, because so far the largest piece of "meat" that has been produced in Eindhoven measured 8 millimeters long, 2 millimeters wide, and 400 microns thick. It contained millions of cells but was about the size of a contact lens. The specimen I saw was as visually stimulating as mouse droppings, and if such a substance can be said to look like anything, it looked like a runny egg. How, I wondered, could those blobs ever feed anyone?

Van der Schaft tried to explain. The initial cells are typically taken from a mouse. (The Dutch have also focused on pork stem cells, because pigs are readily available to them, often reclaimed from eggs discarded at slaughterhouses or taken from biopsies.) Researchers then submerge those cells in amino acids, sugars, and minerals. Generally, that mixture consists of fetal serum taken from calves. Some vegetarians would object even to using two animal cells, and the fetal calf serum would present a bigger problem still. Partly for those reasons, a team working under Klaas Hellingwerf, a microbial physiologist at the University of Amsterdam, has been developing a different growth medium, one based on algae.

After the cells age, van der Schaft and her colleagues place them on biodegradable scaffolds, which help them grow together into muscle tissue. That tissue can then be fused and formed into meat that can be processed as if it were ground beef or pork.

The research is not theoretical, but at this point the Dutch scientists are far more interested in proving that the process will work than in growing meat in commercial quantities. They are preoccupied, in other words, with learning how to make those lens-size blobs more efficiently—not with turning them into hamburgers or meatballs. Great scientists attempt to change the way we think about the natural world but are less concerned with practicalities. They look upon any less fundamental achievement as "an engineering problem," dull but necessary grunt work. "Scientists hate this type of work, because they want breakthroughs, discoveries," Mironov told me. "This is development, not research. And that is the biggest problem we face."

The Dutch team has been trying to discover how best to work with embryonic stem cells, because their flexibility makes them particularly attractive. Stem cells can multiply so quickly that even a few could eventually produce tons of meat. Yet any culture nutritious enough to feed stem cells will have the same effect on bacteria or fungi—both of which grow much more rapidly. "We need completely sterile conditions," van der Schaft explained. "If you accidentally add a single bacterium to a flask, it will be full in one day." There is also the cancer syndrome: stem cells proliferate rapidly and could divide forever if they are maintained properly. That's why they are so valuable. Yet when a cell divides too often it can introduce errors into its genetic code, and these create chromosomal aberrations that can lead to cancer. Tissue engineers need to keep the cells dividing rapidly enough to grow meat on an industrial scale, but not so fast that they become genetic miscreants.

Any group that intends to sell laboratory meat will need to build bioreactors—factories that can grow cells under pristine conditions. Bioreactors aren't new; beer and yeast are made using similar methods. Still, a "carnery," as Nicholas Genovese, the PETA-supported postdoctoral researcher, has suggested such a factory be called, will need much more careful monitoring than a brewery. Muscle cells growing in a laboratory will clump together into a larger version of the gooey mess I had just seen if they're left on their own. To become muscle fibers, the cells have to grow together in an orderly way. Without blood vessels or arteries, there would be no way to deliver oxygen to muscle cells. And without oxygen or nutrients they would starve.

It turns out that muscle cells also need stimulation, because muscles, whether grown in a dish or attached to the biceps of a weight lifter, need to be used or they will atrophy. Tissue fabricated in labs would have to be stimulated with electrical currents. That happens every day in research facilities like the one at Eindhoven; it is not a difficult task with a piece of flesh the size of a fish egg. But to exercise thousands of pounds of meat with electrical currents could potentially cost more than it's worth.

Technical complexities like these have caused some people to suggest that the field will fizzle before one hamburger is sold. Robert Dennis, a professor of biomedical engineering at the University of North Carolina in Chapel Hill, said that the differences between animal tissue and laboratory-created organs remain significant. "Muscle precursor cells grown in a gelatinous scaffold are really just steak-flavored Jell-O," he said. "To reach something that would have real consumer appeal would require stepping back and approaching the question from a fundamentally new direction." Dennis is no less eager to grow meat than his colleagues. He is, however, concerned about hype and false hope. "Engineering fully functional tissues from cells in a petri dish is a monumental technical challenge in terms of both difficulty and long-term impact," he said. "It is right up there with the Apollo program; a permanent and sustainable solution to the global energy and food challenges, appreciated by the public but not yet solved; the global freshwater problem, not yet appreciated by the general public; and global climate change, still vehemently denied by the scientific illiterati. Tissue engineering is well worth the investment, because it will profoundly improve the human condition."

Most others engaged in the research say that the goal isn't quite so distant. "There are many practical difficulties that lie ahead," Frank Baaijens told me. Baaijens is a professor at Eindhoven and a leader in the development of cardiovascular tissue. "But they are not fundamental problems. We know how to do most of what we need to do to make ground meat. We need to learn how to scale it all up. I don't think that is a trivial problem, but industries do this sort of thing all the time. What is needed is the money and the will." Baaijens agreed to work on the project only because it was similar to his current research on the debilitating bedsores that occur when sustained pressure cuts off circulation to vulnerable parts of the body. Without adequate blood flow, the affected tissue dies. "This guy approached us and said, 'You ought to make meat,'" Baaijens recalled. The guy was Willem van Eelen. "We had some doubts, because we were focused on medicine. But he was so enthusiastic and persistent, and in the end I think he was right. We don't necessarily think of this as medicine, but it has the potential to be as valuable as any drug."

Stone Barns, a nonprofit farm in Pocantico Hills, north of New York City, is an eighty-acre agricultural wonderland. The animals and plants there rely on each other to provide food, manure, nutrients, and the symbiotic diversity that any sustainable farm requires. I had come to discuss the future of meat with Dan Barber, the celebrity chef at Blue Hill, the culinary centerpiece of the property. Barber has strong views about the future of agriculture, but he disdains the partisan and evangelical approach so often adopted by food activists. He believes that organic farming can provide solutions to both agricultural and ecological problems. He is not willfully blind, however, to the irony of a farmer in the rich world who thinks that way. "To sit in some of the best farming land in America and talk about what organic food could do to solve the problems of nine hundred million people who go to bed hungry every night . . ." He stopped and smiled wanly. "That is really a pretty good definition of elitist."

When I called a few days earlier and told him that I wanted to talk about lab-grown meat, there was silence on the phone. Then laughter. "Well," he said, "I would rather eat a test-tube hamburger than a Perdue chicken. At least with the burger you are going to know the ingredients." Barber said that he would be perfectly willing to taste such a product. Unlike some other environmentalists, however, he was leery about the ecological value. "If we were replacing some factory-farmed animals, then I suppose it could be used as a complement to agriculture. But removing animals from a good ecological farming system is not beneficial." Barber argues that the vast systems of factory farms in the United States rely on almost limitless supplies of clean water and free energy, which permits farmers to avoid paying a fair price for the carbon used to raise livestock and move their products around the country. Eventually that will have to change, he says, and, when it does, so will the economics of our entire farm system.

It was the first fresh day of spring, and we went out to watch the heritage sows forage in the natural wilds of the farm. They seemed as happy as any person who had just emerged into sunlight from a particularly difficult winter. "The residual benefits of a natural system like this are cultural," Barber said. "These animals are part of a system in which everything is connected. That is why you have to look at the entire life cycle of farms and animals when talking about greenhouse gases."

Barber disputes the common assertion that livestock eating grass belch huge amounts of methane into the atmosphere and are therefore environmentally unacceptable. "That is a simplistic way to look at this problem," he said. "In nature, you just cannot measure methane and say that livestock contribute that amount to climate change and it is therefore a good idea to get rid of livestock. Look at meat. I am not talking about factory farms—which are terrible—or the need for better sources of protein for many people in the world. But if you just look at meat without looking at the life of a cow you are looking at nothing. Cows increase the diversity and resilience of the grass. That helps biological activity in the soil and that helps trap CO_2 from the air. Great soil does that. So when you feed a less methane-emitting animal grain instead of grass you are tying up huge ecosystems into monoculture and plowing and sending enormous amounts of CO_2 into the air with the plows. You are also weakening soil structures that might not come back for hundreds of thousands of years." Stressing that he understood that a growing population will need additional sources of protein, he continued, "So if you can supplement a farming system with cultured meat, that is one thing. But if your goal is to improve animal welfare, ecological integrity, and human health, then replacing animals with laboratory products is the wrong way to go."

The moral and ethical issues that would accompany the use of lab-grown beef may ultimately prove more intractable than the scientific issues. In 2008, when PETA announced a million-dollar reward for the first team to make in vitro chicken, many animal-welfare activists responded with outrage. Jim Thomas, of the environmental group ETC, expressed a common fear: "If test-tube meat hits the big time, we will likely know by its appearance in a Big Mac or when agribusiness buys out the patent holder." Even some PETA leaders felt that the decision to support research into in vitro meat was dangerous. Lisa Lange, a PETA senior vice president, opposed the award. "My main concern is it's our job, as the largest animal rights organization in the world, to introduce the philosophy and hammer it home that animals are not ours to eat," she said.

I can understand why a chef and farmer like Dan Barber believes that we ought to raise animals, kill them humanely, and eat them. I had trouble, though, comprehending why animal rights supporters weren't rushing to embrace a plan that could ultimately end the use of livestock. I put the question to Peter Singer, the Princeton philosopher, who in 1975 published *Animal Liberation,* which is often considered the founding document of the animal rights movement. Singer doesn't come at animal welfare issues from the perspective of a pet owner; he is a utilitarian and believes that it is our moral duty to reduce the amount of suffering on Earth. Since our taste for meat is the only reason that animals are butchered inhumanely and raised in monstrous conditions, he considers eating meat immoral, because it greatly increases the amount of suffering in the world. "It seems all pluses and no minuses to me," he said of in vitro meat. "But I think some vegetarians and vegans just have a 'yuck' response to meat, whatever its source. Or they think it is unhealthy. Or they think that if we accept it, people will think that 'the real thing' is better, whereas we have been trying to tell them for years that it isn't. These are all confusions, in my view. Catholics for centuries taught that masturbation is wrong because sex should lead to procreation. Then IVF comes along, and masturbation is the obvious way to get the sperm that enables an infertile couple to have a child. And the same Catholics say no, masturbation is wrong."

Nobody can yet say whether in vitro meat would find a market. That will depend on the cost and whether people regard it as safe, healthy, and morally acceptable (or perhaps superior) to what we eat today. The last issue is difficult to address. Americans are big fans of the Food Network and of cooking shows such as *Top Chef.* They are eager to follow recipes too. "I wonder how people would feel if, at the beginning of a show, the stars pulled a darling little lamb onto the stage and then beheaded, gutted, and skinned it,"

Ingrid Newkirk said. "I am thinking that the ratings would fall." It may take a sight that shocking for people to fully understand what is at stake. More than anything else, more even than the technical aspects of the science, the success of laboratory-bred meat will depend on our understanding of its importance.

"When I was a kid, I liked to read about science," Bernard Roelen told me. Roelen, a member of the in-vitro-meat research team in the Netherlands, is a stem-cell biologist at Utrecht University. "And often you would read about some problem—nuclear waste, for instance—where it would say we don't know now what the solution is, but scientists will find a solution. And now I am a scientist and we face a really serious issue in the environment. And I feel a responsibility to find an answer.

"Because who will find solutions to these problems? It has to be scientists. We made a mess, and we have to clean it up. I know Willem van Eelen wants to see this happen overnight, and that is not possible. But it will happen eventually, and when it does I think we will look back and wonder why it took so long—why it took so long for us to understand what we have done to animals and to the Earth."

MARK MCCLUSKY

Mad Science

FROM *Wired*

THE PERFECT FRENCH FRY—golden brown, surpassingly crispy on the outside, with a light and fluffy interior that tastes intensely of potato—is not easy to cook. Here's how most people do it at home: cut some potatoes into fry shapes—classic ⅜-inch batons—and toss them into 375-degree oil until they're golden brown. This is a mediocre fry. The center will be raw.

Here's how most restaurants do it: dunk the potatoes in oil twice, once at 325 degrees for about four minutes, until they're cooked through, and then again at 375 degrees to brown them. This is a pretty great fry.

But let's get serious. The chef Heston Blumenthal—owner of the Fat Duck restaurant in Bray, England, holder of three Michelin stars—created what he calls triple-cooked chips. (He's English.) The raw batons are simmered in water until they almost fall apart and then placed on a wire rack inside a vacuum machine that pulls out the moisture. The batons then get the traditional double fry. You need an hour and a $2,000 vacuum chamber, but these are the best fries in the world. Or, rather, they used to be.

The new contender was created by Nathan Myhrvold, the former chief technology officer of Microsoft. Myhrvold cuts his potatoes into batons and rinses them to get rid of surface starch. Then he vacuum-seals them in a plastic bag, in one even layer, with water. He heats the bag to 212 degrees for fifteen minutes, steaming the batons. Then he hits the bag with ultrasound to cavitate the water—forty-five minutes on each side. He reheats the bag in an oven to 212 degrees for five minutes, puts the hot fries on a rack in a vacuum chamber, and then blanches them in 338-degree oil for three minutes. When they're cool, Myhrvold deep-fries the potatoes in oil at 375 degrees until they're crisp, about three more minutes, and then drains them on paper towels. Total preparation time: two hours.

The result is amazing. The outside nearly shatters when you bite into it, yielding to a creamy center that's perfectly smooth. The key is the cavitation caused by the ultrasonic bath—it creates thousands of tiny fissures on the potato's surface, all of which become crunchy when it's fried. When Plato saw the shadow of a french fry on the wall of his cave, the guy standing behind him was snacking on these.

The recipe is one of 1,600 in Myhrvold's new cookbook, *Modernist Cuisine.* It's a big book—2,400 pages big. Six volumes big. Big as in the original slipcase failed Amazon.com's shipping tests and had to be replaced with acrylic. Big like it weighs nearly fifty pounds and costs $625.

This is the way Myhrvold operates. After leaving Microsoft with all the money in the world, he started a company called, immodestly, Intellectual Ventures and turned his attention to busting some of the biggest problems in science and technology. And he dove into a few hobbies. Now most of us, if we were to get interested in cooking, might start to putter around the kitchen at home or do a little reading. Maybe we'd take a class. Because cooking is primarily a craft, dominated by artisans—or artists, if that's how you view what a chef does. Every once in a while, a chemist drops in to take a look or heads for the world of industrial-scale food.

But Myhrvold—a theoretical physicist and computer scientist—has the lifestyle flexibility of a multimillionaire and the mental discipline of a world-class researcher. To him, cooking is about fundamental interactions in the material world: how heat enters food. How you mix two separate materials most effectively. How water molecules interact in a solution. You see a pork chop and some mashed potatoes; he sees a mesh of proteins that coagulate at a specific temperature next to an emulsion of starch and fat. "Chefs think about what it's like to make food," Myhrvold says. "Being a scientist in the kitchen is about asking why something works, and how it works." To him, a kitchen is really just a laboratory that everyone has in his house. And when you have that attitude with that brain and those resources, well, you might not be the best cook in the world, but you just might put together the best cookbook.

If *Modernist Cuisine* lives up to Myhrvold's hopes when it's published in March, it'll be the definitive book about the science of cooking—the *Principia* of the kitchen. It's dense and beautiful and inspired, and even though Myhrvold assembled a team of fifty chefs, writers, photographers, designers, scientists, and editors to create it, the final product is in fact an eerily accurate recapitulation of how Nathan Myhrvold thinks.

Which is to say, the man thinks big about nearly everything. And he wants his french fries to be perfect.

Modernist Cuisine started with a problem.

In 2003 Myhrvold was building his dream house on the shore of Lake Washington outside Seattle, stocking it with esoteric kitchen equipment. One of the toys was a temperature-controlled water bath used for a technique known as sous vide: vacuum-sealing food in plastic bags and cooking it for a long time at relatively low temperatures. Done correctly, it lets a chef precisely control the temperature of the food, so the final product comes out perfect every time.

Myhrvold had come across the technique while studying cooking in France, but he needed information on how long to cook various foods and at what temperature.

And that was the problem. There *wasn't* any information.

For Myhrvold, that's not acceptable. He's a creature of knowledge; talking to him is like taking a graduate seminar. Actually, it's like taking every graduate seminar at once. He bounces from topic to topic as if someone were clicking the remote control through five hundred channels of really high-end BBC documentaries. Here's a lunchtime conversation, only slightly edited:

"Alaska has had more than ten times the number of botulism cases of New York State. But its population is a few percent of New York state. It's because they eat a lot of crap up there. . . . The most thermally diffusive thing that heat travels fastest in is diamond, by a big margin. . . . Suppose you have a broiler with a bunch of separate rods. Turns out there's an optimal distance away from them to have the most even heat. And it's forty-four percent of the distance between them plus five millimeters. . . . The big innovation in the twentieth century wasn't in high-end food, it was in industrial food. . . . Our Carolina barbecue sauces are very thin. We made them authentic thickness. But then we have a note that says two-tenths of a percent of xanthan gum will give you something that clings to your meat and makes your shirt less dirty. Ba dum ba dum ba dum."

That's how Myhrvold cuts off a lot of his own sentences, with what sounds like a kettledrum sound effect for a cartoon somersault. It's not an ellipsis; it's more like his brain has accelerated past the rest of the information. The proof is left as an exercise for the student.

After finding only a couple of articles and one book (in Spanish) about sous vide, Myhrvold posted a message on the high-end culinary discussion forum eGullet asking for sources, recipes, anything. "I sort of naively thought that sous vide was well understood," Myhrvold says. "You heard about people using it, so I figured they clearly must understand it. Well, I discovered that they didn't."

He was no stranger to kitchens. Growing up in Santa Monica, California, with his mother, a model and schoolteacher, Myhrvold started cooking at an early age, checking out cookbooks from the library and preparing elaborate Thanksgiving meals when he was nine. While at Microsoft, he moonlighted in the kitchen of a leading French restaurant in Seattle for nearly two years.

But he's primarily a scientist. Myhrvold has a master's degree in geophysics and space physics and another one in mathematical economics. He got his PhD in theoretical and mathematical physics from Princeton at twenty-three and did a postdoctoral fellowship with Stephen Hawking at Cambridge. He started a software company that Microsoft bought in 1986, founded Microsoft Research in 1991, and left the company as its CTO and chief strategist in 1999. He has hundreds of patents issued or pending. Oh, and he's also a photographer, a patron of paleontology research, and a world-champion barbecue chef. Seriously.

So Myhrvold the cook and Myhrvold the scientist went to work. Chicken and salmon and beef all got sous vided, with temperature probes inserted so Myhrvold could track how the heat moved through the food. He wrote a program using Mathematica to model the heat transfer through various shapes and sizes of food without actually having to cook. "I got kind of carried away," he says.

Almost a year and a half after asking his question on eGullet, Myhrvold answered it himself, posting the results of his experiments—charts that showed how long and at what temperature to cook a certain piece of food to get to a desired final temperature. Instantly the thread became the definitive reference to sous vide.

By the time someone online suggested that he write a book based on the information, Myhrvold had already moved on to looking at food safety concerns raised by the low temperatures used in sous vide. He was even helping chefs convince food inspectors that the technique was safe. "From there I sort of decided, hey, why not do the whole thing?" Myhrvold says. "It made sense at the time."

The Intellectual Ventures Lab, hard by a tennis practice facility and an auto-repair shop on the outskirts of Bellevue, Washington, isn't just easy to miss—it's almost as though it was scientifically designed to look as nondescript as possible. Inside the former Harley-Davidson garage, though, is 27,500 square feet of thinking space—as much a physical manifestation of Myhrvold's polymath mind as the cookbook is a literary and photographic one. Just inside the front door are the wet chemistry lab, the physics lab, the repair shop, and the laser testing rooms. A space farther back and to the right is crammed with computer-controlled milling machines that carve objects from metal or plastic with millimeter precision and a giant water-jet cutting table. It's hundreds of thousands of bucks worth of gear—a factory for fabricating anything a scientist might need.

Inside, dozens of PhDs work on a bevy of projects. One group is trying to perfect an idea that scientists have been hammering on since the 1950s—a traveling-wave nuclear reactor. It could, in theory, run for fifty to a hundred years without needing to be refueled, primarily on uranium 238, which is a cheap, nonweaponizable byproduct of the uranium-enrichment process. The TerraPower project, as it's called, should yield a prototype reactor by 2020.

Then there's the Salter Sink, which is supposed to lessen the impact of hurricanes by funneling warm water from the ocean's surface into the colder water below. And there's the solution that the company has proposed to slow global warming: pump sulfur dioxide into the stratosphere to mirror the cooling effect caused by large volcanic eruptions. Al Gore told the authors of *SuperFreakonomics* that the plan was "nuts," but that's of little consequence to Myhrvold; several Nobel laureates agree with him that the sulfur scheme might work.

Scientists at Intellectual Ventures have invented a new x-ray scanner that produces clearer images, surfaces that sterilize themselves, a portable freezer that keeps vaccines active without electricity, and even metamaterials that could reverse light, creating a cloaking device. But perhaps the flashiest creation from the lab is a bug zapper called the photonic fence, which Myhrvold unveiled at the TED Conference in 2010. It's a result of the company's ongoing work to eliminate malaria in response to a challenge from Myhrvold's old boss, Bill Gates. At a brainstorming session in 2008, someone suggested that lasers could kill mosquitoes before they could spread the disease—a kind of insect-world Star Wars laser-defense system.

The team pulled together parts from consumer electronics and eBay to develop a prototype—one that could even determine if a mosquito was male or female (only the females bite humans). The females would be blasted out of the sky; the males would be left alone. It's a massively clever bit of engineering and coding—and the parts are cheap and getting cheaper. Like the other gadgets, it's so crazy that it just might work.

Malaria is the focus of a lot of effort at the lab. Across the street in the annex (a former interior design showroom that still has some cabinet display models on the walls) is the company's supercomputer, built from 1,000 Xeon core processors, which the mathematician Philip Eckhoff is using to model the spread and potential eradication of the disease.

Almost all of these inventions have one thing in common: Intellectual Ventures doesn't want to manufacture them. The company's business is making money from ideas, not from the products that the ideas could generate.

Myhrvold says that the company is trying to create a capital market for inventions, a market for intangible intellectual property like the one that grew around software in the 1980s. But Intellectual Ventures isn't just doing its own research and brainstorming leading to patents, it is also (much more controversially) buying up thousands of patents from outside inventors, which it then licenses to technology companies like Apple, Google, and Sony.

To Myhrvold, this is an elegant, scientifically minded hack of the patent system—where people can patent not only products but ideas. To intellectual-property purists, though, that sort of behavior is called patent trolling—gathering the rights to ideas and then forcing companies to pay up when those ideas actually appear in the world and are about to be turned into usable technology. And indeed, in December, Intellectual Ventures filed three lawsuits claiming that nine companies were infringing some of its patents.

But the accusation of trolling has become increasingly frustrating to Myhrvold. "If you look at the list of people who have been called patent trolls," he says, "it's everyone who's ever filed a patent suit." He points out that his company applies for patents on five hundred of its own inventions every year. And anyway, he says, the system is designed for this kind of transaction. "Some people think it's scandalous. 'Oh my gosh, they buy patents!' Well, yeah. And publishers buy books from writers," Myhrvold says. "I've never gotten it, except that there are people who have ideological—bordering on religious—ideas about intellectual property, most of which are in my view not very deeply thought through."

That's Myhrvold. On one hand, there's the fevered imagination and brainstorming, invention and science, the quest to change the world. But on the other hand, there's the aggressive businessman who isn't just around to create cool stuff—he's looking to make a ton of money, too.

Writing about sous vide led Myhrvold to think more deeply about how heat moves through different media (which is why *Modernist Cuisine* may well be the only cookbook ever published with a long disquisition on Fourier's law, the equation for calculating heat transfer). That led to food safety, and that led to a more general exploration of the microbiology of food. Myhrvold soon realized that his ambition for *Modernist Cuisine* had outstripped his ability to write it alone. "It's like writing software," he says. "If you want to do interesting software, you have to have a bunch of people do it, because the amount of software that one person can do isn't that interesting."

A chef would have built a kitchen; Myhrvold built the Cooking Lab. He carved out a corner of the Intellectual Ventures lab and filled it with gear—not just stoves and ovens but industrial-grade homogenizers, freeze-driers, steam-heated ovens, and vacuum distillation machines. If Thomas Edison and Martha Stewart built a house, this is what the kitchen would look like.

And then, like the primary investigator in an academic laboratory, Myhrvold started hiring researchers. He began with Chris Young, a thirty-four-year-old with degrees in math and biochemistry from the University of Washington and one of the plummiest jobs in cooking, running the development kitchen at Blumenthal's Fat Duck. But in 2007, he was ready to come home. Five years in the town of Bray, thirty miles west of London, was enough for Young; he was set to move to a job at a San Francisco Bay Area restaurant when he e-mailed Myhrvold, with whom he had corresponded about food science, to give him his new coordinates.

Three minutes later, a message from Myhrvold appeared on Young's screen. It had the subject line "Crazy Idea." The note was one line long: "Why don't you come work for me?"

Young signed up and brought in Maxime Bilet, a young chef he had worked with at the Fat Duck, to run the kitchen day to day. Wayt Gibbs, a former editor at *Scientific American* who works at Intellectual Ventures, was drafted to handle the editing, while photographer Ryan Matthew Smith joined the team after responding to a craigslist job posting.

Myhrvold then let them explore largely on their own. "Nathan creates a dynamic, free-thinking environment here," Young says. "This is a unique place to work. You'll be in the kitchen, and then someone like Neal Stephenson will wander by." For example, when the chefs were working on the part of the book focused on gels and thickeners, Myhrvold was having them concentrate on exotic hydrocolloids like agar or gellan. But then Bilet and the culinary team came to him with a suggestion. They wanted to add egg gels—custards, basically. "They're just as valid," Bilet told Myhrvold. "They're just as cool." Myhrvold gave them the green light, and the team hit their library of hundreds of food-science books to see what people already knew about eggs and how they cook.

Then they started collecting data, cooking hundreds of batches of egg custard. Each time they tweaked a variable—temperature, yolk-to-egg-white ratio, amount of liquids.

It took them two weeks, all for a deceptively simple chart. Temperatures are on one axis and the ratio of egg to liquid is on the other; cross-reference the two and you can choose a texture, from a runny crème anglaise to a firm flan. "All that work and it condenses down into this one little teeny table," Myhrvold says. Of course, that table is an unprecedented master course in egg cookery. "It's really cool to be in an experimental kitchen like this," Young says. "If you need to, you can talk to an engineer or a physicist. We have access to all of their analytical tools, and if our equipment breaks, we have these PhDs here to help us fix it. It's just really eclectic."

Working next to all those other projects has required a few adjustments, however. One night, the cookbook team was in the kitchen late, testing new recipes. The photonic fence team was also working late, seeing if their tracking software could follow mosquitoes at long distances. They had put a box filled with bugs at the top of a set of stairs at one end of the hundred-foot-long Cooking Lab and set up their laser at the other end. As the chefs stood at their stoves, the beams started flashing above them. "I guarantee that we are the only kitchen in the world that had lasers overhead," Young says. "They told me they were firing at a nonkilling intensity."

Lunch at the cooking lab. First comes raspberry gazpacho with piquillo peppers and macadamia nuts. Foie gras and horse mackerel are served with sous vide ponzu. Mushroom omelets are cooked in a steam oven, keeping them moist and tender. Comte cheese is turned into an aerated sponge with a vacuum machine and is served with hazelnut cakes. It's twelve courses overall, each one highlighting a different cutting-edge tool or technique.

When spot prawns and carotene butter show up—cook carrots in butter and then separate out the solids with a centrifuge—Myhrvold takes a bite, thinks for a moment, and then asks Bilet to hang on a second. "This is great, Max," Myhrvold says. "You know what I think it needs? It needs something crunchy."

"A little texture?" Bilet asks.

"Yeah. How about some freeze-dried carrot? Little chunks."

Bilet hesitates, looking at the dish. He seems dubious.

"Either that or something else crunchy," Myhrvold says. "Because it's fantastic but could use a texture element."

"We could do something with coconut," Bilet suggests. "To balance the carrot. Maybe a savory coconut tuile with freeze-dried prawn powder."

"That would do it," Myhrvold says. He goes back to eating.

Myhrvold is not a professional chef, but he's turned himself into a professional eater—thousands of hours of culinary training and meals at hundreds of the world's best restaurants. He's a scientific Falstaff, a rare combination of rationalist and sensualist. In fact, lunch at the lab would stand up to the food at some of the most avant-garde restaurants in the world. That's an abiding passion of Myhrvold's, right there in the title of the book. For him, there's nothing in the food world more exciting than the science-driven cooking he calls modernist.

Over the past two decades, a wave of chefs—Blumenthal, Grant Achatz of Chicago's Alinea, and Ferran Adrià, whose restaurant El Bulli in Spain is considered one of the world's best—have looked beyond tradition for ways to manipulate their food. Adrià uses everything from industrial food additives to freeze-drying in the pursuit of otherworldly effects, like a soup that changes temperature in your mouth as you eat it. It's what some writers (though not the chefs themselves) have called "molecular gastronomy," and a major thrust of *Modernist Cuisine* is to explain just what the hell is going on in these high-end kitchens. "There's a set of cooking techniques that go back twenty-some years that are hugely interesting to people, very useful, poorly understood, and almost impossible to learn," Myhrvold says. "The best you can do is to go cook at a few restaurants that do this, and you come away with like one percent of it."

Until now there's been no comprehensive reference. And the ingredients require a precision unfamiliar to many cooks. As Myhrvold observes, "It's a superbad idea to put a 'pinch' of xanthan gum in something." *Modernist Cuisine* sets out to explain and expand the chef's toolkit. "One of the wonderful things about the book is that it makes it clear what these things are good for, what they're not good for, what their strengths and weaknesses are," says Harold McGee, the author of the seminal food-science book *On Food and Cooking*. "I think it will go a long way toward demystifying and also expanding the number of people who can play with them and come up with new things."

That's a big change from cooking's artisanal roots. "You were taught how to make a hollandaise sauce, and you were never really taught why it works," says Thomas Keller, who runs Per Se in New York City and the French Laundry in northern California and is generally considered the best chef in the United States. "You were just taught how to make it, and you were taught how to fix it if it broke, and that was it." Myhrvold and his team want cooks to understand the science behind the technique. So *Modernist Cuisine* explains the avant-garde by emphasizing the most basic elements of cooking: heat and water.

Myhrvold has a favorite riddle: "If you have two steaks, one that's an inch thick, one that's two inches thick, how much longer does the thicker one need to cook?"

If you said the thicker steak takes twice as long, you're making the same mistake most cooks do. "It's four times as long. It goes roughly like the square," Myhrvold says. "How come cookbooks don't tell you that?" he asks, nearly bursting with indignation. The fundamental laws of heat, he figures, are the fundamental laws of cooking. "The physics of heat is diffusion," Myhrvold says. "So that's also the physics of drying things or of marination. They're all about diffusing things. The physics of heating things is also the physics of cooling things. It's the same basics over and over."

Then there's water. "Three things about water affect almost all of cooking," Myhrvold says. "First are the hydrogen bonds, which is why it has an incredibly high boiling point. Another is that it's a polar molecule, so that it dissolves a lot of things, and there are things that won't mix with it. And then there's how much energy it takes to heat water. That's why steaming food works; that's why pressure cookers work."

This isn't like most writing about food science. McGee's science-minded *On Food and Cooking* is a de facto reference in every professional kitchen—and many amateur ones. McGee says he's a fan of Myhrvold's work; the two men are friends, in fact. "I'm much more interested in the chemistry of flavor than Nathan is," McGee says. "That has to do with the diversity of compounds that you find in nature, how they get there, and how we detect them."

It's a polite sort of turf-carving, and Myhrvold is in just as much of a rush to establish his own. "In terms of broadly looking at food science and chemistry, and trying to explain it to a lay audience,

Harold led the way," Myhrvold says. "But we have a physics-oriented book." Most cooks focus on the difference between filet mignon and rib eye. Myhrvold and his team want you to comprehend the whole cow. "If all you want to do is follow recipes, you don't need insights," he says. "If you want to do new things, you have to understand what the hell you're doing."

The ambition, the sheer bigness of *Modernist Cuisine,* does trigger the oh-come-on meter just a bit. Saying cooks need to understand the physics of diffusion is a little bit like saying a home woodworker needs to understand quantum mechanics. Sure, Planck's constant helps explain how nails go through maple, but calculating the one doesn't help you hammer the other.

Ironically, *Modernist Cuisine* will start tormenting UPS drivers with its bulk at the same time that the movement it celebrates—avant-garde, science-driven cooking—is waning. Ferran Adrià is closing El Bulli this year. Achatz is opening a new restaurant this spring that won't emphasize the techniques he helped popularize. "I think the book will have long-lasting importance in gastronomy," Achatz says. "But the particular style of cooking that it highlights might not. It's clear that the tide is turning. I don't think many chefs will continue to take the wholehearted scientific approach."

The tools and techniques that chefs like Adrià and Achatz popularized are trickling down. Flip on *Top Chef* and you're likely to see someone mucking about with liquid nitrogen and vacuum sealers. But the artistic part, the creativity of avant-garde chefs that Myhrvold finds so inspiring, seems to be shrinking. If that's so, *Modernist Cuisine* isn't the *Principia* of the kitchen but its *Consolation of Philosophy,* the book that collects and summarizes all the knowledge in a field at the moment the field implodes. It's a eulogy.

Myhrvold has certainly considered that possibility. To get this book out, he spent hundreds of thousands, maybe even millions, hiring staff, building a lab, setting up a separate company to self-publish it. And while he might be relatively immune to financial pressures, there's another judgment that the market will make. "One of the names for small-volume personal publishing is vanity publishing," Myhrvold says. "So is this useful to people, or is it entirely vanity? That's a fascinating question. If no one wants it, you have to ask yourself, what am I doing it for?"

It's almost impossible to comprehend all of *Modernist Cuisine*. It seeks to be the first and last word in its field, to settle every argument, to capture all of human knowledge about cooking. And, ultimately, it's a book that utterly reflects Myhrvold. "We had a focus on physics. We had a focus on computer modeling. We had a focus on photography," Myhrvold says. "Those are all things that I'm completely into. We had a focus on the history and the philosophy of this kind of cuisine. Again, that's totally what I'm into." That's why the criticisms won't matter too much to Myhrvold. In the end, *Modernist Cuisine* is more than a cookbook. It's an autobiography—the world's most oblique memoir, so accurate a reflection of its creator that he might be the only person in the world who fully understands it.

RIVKA GALCHEN

Dream Machine

FROM *The New Yorker*

ON THE OUTSKIRTS of Oxford lives a brilliant and distressingly thin physicist named David Deutsch, who believes in multiple universes and has conceived of an as yet unbuildable computer to test their existence. His books have titles of colossal confidence *(The Fabric of Reality, The Beginning of Infinity).* He rarely leaves his house. Many of his close colleagues haven't seen him for years, except at occasional conferences via Skype.

Deutsch, who has never held a job, is essentially the founding father of quantum computing, a field that devises distinctly powerful computers based on the branch of physics known as quantum mechanics. With one-millionth of the hardware of an ordinary laptop, a quantum computer could store as many bits of information as there are particles in the universe. It could break previously unbreakable codes. It could answer questions about quantum mechanics that are currently far too complicated for a regular computer to handle. None of which is to say that anyone yet knows what we would really do with one. Ask a physicist what, practically, a quantum computer would be "good for," and he might tell the story of the nineteenth-century English scientist Michael Faraday, a seminal figure in the field of electromagnetism, who, when asked how an electromagnetic effect could be useful, answered that he didn't know, but he was sure that one day it could be taxed by the queen.

In a stairwell of Oxford's Clarendon Physics Laboratory there is a photo poster from the late 1990s commemorating the Oxford Center for Quantum Computation. The photograph shows a well-groomed crowd of physicists gathered on the lawn. Photoshopped into a far corner, with the shadows all wrong, is the head of David Deutsch, looking like a time traveler teleported in for the day. It is tempting to interpret Deutsch's representation in the photograph as a collegial joke, because of Deutsch's belief that if a quantum computer were built it would constitute near-irrefutable evidence of what is known as the Many Worlds Interpretation of quantum mechanics, a theory that proposes pretty much what one would imagine it does. A number of respected thinkers in physics besides Deutsch support the Many Worlds Interpretation, though they are a minority, and primarily educated in England, where the intense interest in quantum computing has at times been termed the Oxford flu.

But the infection of Deutsch's thinking has mutated and gone pandemic. Other scientists, although generally indifferent to the truth or falsehood of Many Worlds as a description of the universe, are now working to build these dreamed-up quantum-computing machines. Researchers at centers in Singapore, Canada, and New Haven, in collaboration with groups such as Google and NASA, may soon build machines that will make today's computers look like pocket calculators. But Deutsch complements the indifference of his colleagues to Many Worlds with one of his own—a professional indifference to the actual building of a quantum computer.

Physics advances by accepting absurdities. Its history is one of unbelievable ideas proving to be true. Aristotle quite reasonably thought that an object in motion, left alone, would eventually come to rest; Newton discovered that this wasn't true, and from there worked out the foundation of what we now call classical mechanics. Similarly, physics surprised us with the facts that Earth revolves around the sun, time is curved, and the universe if viewed from the outside is beige.

"Our imagination is stretched to the utmost," the Nobel Prize–winning physicist Richard Feynman noted, "not, as in fiction, to imagine things which are not really there, but just to comprehend those things which *are* there." Physics is strange, and the people who spend their life devoted to its study are more accustomed to its strangeness than the rest of us. But even to physicists, quantum mechanics—the basis of a quantum computer—is almost intolerably odd.

Quantum mechanics describes the natural history of matter and energy making their way through space and time. Classical mechanics does much the same, but while classical mechanics is very accurate when describing most of what we see (sand, baseballs, planets), its descriptions of matter at a smaller scale are simply wrong. At a fine enough resolution, all those reliable rules about balls on inclined planes start to fail.

Quantum mechanics states that particles can be in two places at once, a quality called superposition; that two particles can be related, or "entangled," such that they can instantly coordinate their properties, regardless of their distance apart in space and time; and that when we look at particles we unavoidably alter them. Also, in quantum mechanics the universe, at its most elemental level, is random, an idea that tends to upset people. Confess your confusion about quantum mechanics to a physicist and you will be told not to feel bad, because physicists find it confusing, too. If classical mechanics is George Eliot, quantum mechanics is Kafka.

All the oddness would be easier to tolerate if quantum mechanics merely described marginal bits of matter or energy. But it is the physics of everything. Even Einstein, who felt at ease with the idea of wormholes through time, was so bothered by the whole business that in 1935 he coauthored a paper titled "Can Quantum-Mechanical Description of Physical Reality Be Considered Complete?" He pointed out some of quantum mechanics' strange implications, and then answered his question, essentially, in the negative. Einstein found entanglement particularly troubling, denigrating it as "spooky action at a distance," a telling phrase, which consciously echoed the seventeenth-century disparagement of gravity.

The Danish physicist Niels Bohr took issue with Einstein. He argued that in quantum mechanics, physics had run up against the limit of what science could hope to know. What seemed like nonsense *was* nonsense, and we needed to realize that science, though wonderfully good at predicting the outcomes of individual experiments, could not tell us about reality itself, which would remain forever behind a veil. Science merely revealed what reality looked like to us.

Bohr's stance prevailed over Einstein's. "Of course, both sides of that dispute were wrong," Deutsch observed, "but Bohr was trying to obfuscate, whereas Einstein was actually trying to solve the problem." As Deutsch notes in *The Fabric of Reality,* "To say that prediction is the purpose of a scientific theory is to confuse means with ends. It is like saying that the purpose of a spaceship is to burn fuel." After Bohr, a "shut up and calculate" philosophy took over physics for decades. To delve into quantum mechanics as if its equations told the story of reality itself was considered sadly misguided, like those earnest inquiries people mail to 221B Baker Street, addressed to Sherlock Holmes.

I met David Deutsch at his home, at four o'clock on a wintry Thursday afternoon. Deutsch grew up in the London area, took his undergraduate degree at Cambridge, stayed there for a master's in math—which he claims he's no good at—and went on to Oxford for a doctorate in physics. Though affiliated with the university, he is not on staff and has never taught a course. "I love to give talks," he told me. "I just don't like giving talks that people don't want to hear. It's wrong to set up the educational system that way. But that's not why I don't teach. I don't teach for visceral reasons—I just dislike it. If I were a biologist, I would be a theoretical biologist, because I don't like the idea of cutting up frogs. Not for moral reasons but because it's disgusting. Similarly, talking to a group of people who don't want to be there is disgusting." Instead, Deutsch has made money from lectures, grants, prizes, and his books.

In the half-light of the winter sun, Deutsch's house looked a little shabby. The yard was full of what appeared to be English ivy, and near the entrance was something twiggy and bushlike that was either dormant or dead. A handwritten sign on the door said that deliveries should "knock hard." Deutsch answered the door. "I'm very much in a rush," he told me, before I'd even stepped inside. "In a rush about so many things." His thinness contributed to an oscillation of his apparent age between nineteen and a hundred and nineteen. (He's fifty-seven.) His eyes, behind thick glasses, appeared outsized, like those of an appealing anime character. His vestibule was cluttered with old phone books, cardboard boxes, and piles of papers. "Which isn't to say that I don't have time to talk to you," he continued. "It's just that—that's why the house is in such disarray, because I'm so rushed."

More than one of Deutsch's colleagues told me about a Japanese documentary film crew that had wanted to interview Deutsch at his house. The crew asked if they could clean up the house a bit. Deutsch didn't like the idea, so the film crew promised that after filming they would reconstruct the mess as it was before. They took extensive photographs, like investigators at a crime scene, and then cleaned up. After the interview, the crew carefully reconstructed the former "disorder." Deutsch said he could still find things, which was what he had been worried about.

Taped onto the walls of Deutsch's living room were a map of the world, a periodic table, a hand-drawn cartoon of Karl Popper, a poster of the signing of the Declaration of Independence, a taxonomy of animals, a taxonomy of the characters in *The Simpsons,* color printouts of pictures of McCain and Obama, with handwritten labels reading "this one" and "that one," and two color prints of an actor who looked to me a bit like Hugh Grant. There were also old VHS tapes, an unused fireplace, a stationary exercise bike, and a large flat-screen television whose newness had no visible companion. Deutsch offered me tea and biscuits. I asked him about the Hugh Grant look-alike.

"You obviously don't watch much television," he replied. The man in the photographs was Hugh Laurie, a British actor known for his role in the American medical show *House.* Deutsch described *House* to me as "a great program about epistemology, which, apart from fundamental physics, is really my core interest. It's a program about the myriad ways that knowledge can grow or can fail to grow." Dr. House is based on Sherlock Holmes, Deutsch informed me. "And House has a friend, Wilson, who is based on Watson. Like Holmes, House is an arch-rationalist. Everything's got to have a reason, and if he doesn't know the reason it's because he doesn't know it, not because there isn't one. That's an essential attitude in fundamental science." One imagines the ghost of Bohr would disagree.

Deutsch's reputation as a cloistered genius stems in large part from his foundational work in quantum computing. Since the 1930s, the field of computer science has held on to the idea of a universal computer, a notion first worked out by the field's modern founder, the British polymath Alan Turing. A universal computer would be capable of comporting itself like any other computer, just as a synthesizer can make the sounds made by any other musical instrument. In a 1985 paper, Deutsch pointed out that because Turing was working with classical physics, his universal computer could imitate only a subset of possible computers. Turing's theory needed to account for quantum mechanics if its logic was to hold. Deutsch proposed a universal computer based on quantum physics, which would have calculating powers that Turing's computer (even in theory) could not simulate.

According to Deutsch, the insight for that paper came from a conversation in the early eighties with the physicist Charles Bennett of IBM about computational-complexity theory, at the time a sexy new field that investigated the difficulty of a computational task. Deutsch questioned whether computational complexity was a fundamental or a relative property. Mass, for instance, is a fundamental property, because it remains the same in any setting; weight is a relative property, because an object's weight depends on the strength of gravity acting on it. Identical baseballs on Earth and on the moon have equivalent masses but different weights. If computational complexity was like mass—if it was a fundamental property—then complexity was quite profound; if not, then not.

"I was just sounding off," Deutsch said. "I said they make too much of this"—meaning complexity theory—"because there's no standard computer with respect to which you should be calculating the complexity of the task." Just as an object's weight depends on the force of gravity in which it's measured, the degree of computational complexity depended on the computer on which it was measured. One could find out how complex a task was to perform on a particular computer, but that didn't say how complex a task was *fundamentally,* in reference to the universe. Unless there really was such a thing as a universal computer, there was no way a description of complexity could be fundamental. Complexity theorists, Deutsch reasoned, were wasting their time.

Deutsch continued, "Then Charlie said, quietly, 'Well, the thing is, there *is* a fundamental computer. The fundamental computer is physics itself.'" That impressed Deutsch. Computational complexity was a fundamental property; its value referenced how complicated a computation was on that most universal computer, that of the physics of the world. "I realized that Charlie was right about that," Deutsch said. "Then I thought, But these guys are using the wrong physics. They realized that complexity theory was a statement about physics, but they didn't realize that it mattered whether you used the true laws of physics, or some approximation, i.e., classical physics." Deutsch began rewriting Turing's universal-computer work using quantum physics. "Some of the differences are very large," he said. Thus, at least in Deutsch's mind, the quantum universal computer was born.

A number of physics journals rejected some of Deutsch's early quantum-computing work, saying it was "too philosophical." When it was finally published, he said, "a handful of people kind of got it." One of them was the physicist Artur Ekert, who had come to Oxford as a graduate student and who told me, "David was really the first one who formulated the concept of a quantum computer."

Other important figures early in the field included the reclusive physicist Stephen J. Wiesner, who, with Bennett's encouragement, developed ideas like quantum money (uncounterfeitable!) and quantum cryptography, and the philosopher of physics David Albert, whose imagining of introspective quantum automata (think robots in analysis) Deutsch describes in his 1985 paper as an example of "a true quantum computer." Ekert says of the field, "We're a bunch of odd ducks."

Although Deutsch was not formally Ekert's adviser, Ekert studied with him. "He kind of adopted me," Ekert recalled, "and then, afterward, I kind of adopted him. My tutorials at his place would start at around eight P.M., when David would be having his lunch. We'd stay talking and working until the wee hours of the morning. He likes just talking things over. I would leave at three or four A.M., and then David would start properly working afterward. If we came up with something, we would write the paper, but sometimes we wouldn't write the paper, and if someone else also came up with the solution we'd say, 'Good, now we don't have to write it up.'" It was not yet clear, even in theory, what a quantum computer might be better at than a classical computer, so Deutsch and Ekert tried to develop algorithms for problems that were intractable on a classical computer but that might be tractable on a quantum one.

One such problem is prime factorization. A holy grail of mathematics for centuries, it is the basis of much current cryptography. It's easy to take two large prime numbers and multiply them, but it's very difficult to take a large number that is the product of two primes and then deduce what the original prime factors are. To factor a number of two hundred digits or more would take a regular computer many lifetimes. Prime factorization is an example of a process that is easy one way (easy to scramble eggs) and very difficult the other (nearly impossible to unscramble them). In cryptography, two large prime numbers are multiplied to create a security key. Unlocking that key would be the equivalent of unscrambling an egg. Using prime factorization in this way is called RSA encryption (named for the scientists who proposed it, Rivest, Shamir, and Adleman), and it's how most everything is kept secret on the Internet, from your credit card information to IRS records.

In 1992 the MIT mathematician Peter Shor heard a talk about theoretical quantum computing, which brought to his attention the work of Deutsch and other foundational thinkers in what was then still an obscure field. Shor worked on the factorization problem in private. "I wasn't sure anything would come of it," Shor explained. But about a year later, he emerged with an algorithm that (a) could only be run on a quantum computer and (b) could quickly find the prime factors of a very large number—the grail! With Shor's algorithm, calculations that would take a normal computer longer than the history of the universe would take a sufficiently powerful quantum computer an afternoon. "Shor's work was the biggest jump," the physicist David DiVincenzo, who is considered among the most knowledgeable about the history of quantum computing, says. "It was the moment when we were, like, Oh, now we see what it would be good for."

Today quantum computation has the sustained attention of experimentalists; it also has serious public and private funding. Venture capital companies are already investing in quantum encryption devices, and university research groups around the world have large teams working both to build hardware and to develop quantum-computer applications—for example, to model proteins or to better understand the properties of superconductors.

Artur Ekert became a key figure in the transition from pure theory to building machines. He founded the quantum computation center at Oxford, as well as a similar center a few years later at Cambridge. He now leads a center in Singapore, where the government has made quantum-computing research one of its top goals. "Today in the field there's a lot of focus on lab implementation, on how and from what you could actually build a quantum computer," DiVincenzo said. "From the perspective of just counting, you can say that the majority of the field now is involved in trying to build some hardware. That's a result of the success of the field." In 2009 Google announced that it had been working on quantum-computing algorithms for three years, with the aim of having a computer that could quickly identify particular things or people from among vast stores of video and images—David Deutsch, say, from among millions of untagged photographs.

In the early nineteenth century, a "computer" was any person who computed: someone who did the math for building a bridge, for example. Around 1830, the English mathematician and inventor Charles Babbage worked out the idea for his Analytical Engine, a machine that would remove the human from computing and thus bypass human error. Nearly no one imagined that an analytical engine would be of much use, and in Babbage's time no such machine was ever built to completion. Though Babbage was prone to serious mental breakdowns, and though his bent of mind was so odd that he once wrote to Alfred, Lord Tennyson, correcting his math (Babbage suggested rewriting "Every minute dies a man / Every minute one is born" as "Every moment dies a man / Every moment one and a sixteenth is born," further noting that although the exact figure was 1.167, "something must, of course, be conceded to the laws of meter"), we can now say the guy was on to something.

A classical computer—any computer we know today—transforms an input into an output through nothing more than the manipulation of binary bits, units of information that can be either zero or one. A quantum computer is in many ways like a regular computer, but instead of bits it uses qubits. Each qubit (pronounced "q-bit") can be zero or one, like a bit, but a qubit can also be zero *and* one—the quantum-mechanical quirk known as superposition. It is the state that the cat in the classic example of Schrödinger's closed box is stuck in: dead and alive at the same time. If one reads quantum-mechanical equations literally, superposition is ontological, not epistemological; it's not that we don't *know* which state the cat is in but that the cat really *is* in both states at once. Superposition is like Freud's description of true ambivalence: not feeling unsure, but feeling opposing extremes of conviction at once. And, just as ambivalence holds more information than any single emotion, a qubit holds more information than a bit.

What quantum mechanics calls entanglement also contributes

to the singular powers of qubits. Entangled particles have a kind of ESP: regardless of distance, they can instantly share information that an observer cannot even perceive is there. Input into a quantum computer can thus be dispersed among entangled qubits, which lets the processing of that information be spread out as well: tell one particle something, and it can instantly spread the word among all the other particles with which it's entangled.

There's information that we can't perceive when it's held among entangled particles; that information is their collective secret. As quantum mechanics has taught us, things are inexorably changed by our trying to ascertain anything about them. Once observed, qubits are no longer in a state of entanglement or of superposition: the cat commits irrevocably to life or death, and this ruins the quantum computer's distinct calculating power. A quantum computer is the pot that, if watched, really won't boil. Charles Bennett described quantum information as being "like the information of a dream—we can't show it to others, and when we try to describe it we change the memory of it."

But once the work on the problem has been done among the entangled particles, then we can look. When one turns to a quantum computer for an "answer," that answer, from having been held in that strange entangled way among many particles, needs then to surface in just one ordinary, unentangled place. That transition from entanglement to nonentanglement is sometimes termed "collapse." Once the system has collapsed, the information it holds is no longer a dream or a secret or a strange cat at once alive and dead; the answer is then just an ordinary thing we can read off a screen.

Qubits are not merely theoretical. Early work in quantum-computer hardware built qubits by manipulating the magnetic nuclei of atoms in a liquid soup with electrical impulses. Later teams, such as the one at Oxford, developed qubits using single trapped ions, a method that confines charged atomic particles to a particular space. These qubits are very precise, though delicate; protecting them from interference is quite difficult. More easily manipulated, albeit less precise, qubits have been built from superconducting materials arranged to model an atom. Typically, the fabrication of a qubit is not all that different from that of a regular chip. At Oxford I saw something that resembled an oversize air-hockey table chaotically populated with a specialty Lego set, with what looked like a salad-bar sneeze guard hovering over it; this extended apparatus comprised lasers and magnetic-field generators and optical cavities, all arranged at just the right angles to manipulate and protect from interference the eight tiny qubits housed in a steel tube at the table's center.

Oxford's eight-qubit quantum computer has significantly less computational power than an abacus, but fifty to a hundred qubits could make something as powerful as any laptop. A team in Bristol, England, has a small, four-qubit quantum computer that can factor the number 15. A Canadian company claims to have built one that can do Sudoku, though that has been questioned by some who say that the processing is effectively being done by normal bits, without any superposition or entanglement.

Increasing the number of qubits, and thus the computer's power, is more than a simple matter of stacking. "One of the main problems with scaling up is a qubit's fidelity," Robert Schoelkopf, a physics professor at Yale who leads a quantum-computing team, explained. By fidelity, he refers to the fact that qubits "decohere" —fall out of their information-holding state—very easily. "Right now, qubits can be faithful for about a microsecond. And our calculations take about one hundred nanoseconds. Either calculations need to go faster or qubits need to be made more faithful."

What qubits are doing as we avert our gaze is a matter of some dispute and occasionally—"shut up and calculate"—of some determined indifference, especially for more pragmatically minded physicists. For Deutsch, to really understand the workings of a quantum computer necessitates subscribing to Hugh Everett's Many Worlds Interpretation of quantum mechanics.

Everett's theory was neglected upon its publication in 1957 and is still a minority view. It entails the following counterintuitive reasoning: every time there is more than one possible outcome, all of them occur. So if a radioactive atom might or might not decay at any given second, it both does and doesn't; in one universe it does, and in another it doesn't. These small branchings of possibility then ripple out until everything that is possible in fact *is*. According to Many Worlds theory, instead of a single history there are innumerable branchings. In one universe your cat has died, in another he hasn't, in a third you died in a sledding accident at age seven and never put your cat in the box in the first place, and so on.

Many Worlds is an ontologically extravagant proposition. But it also bears some comfortingly prosaic implications: in Many Worlds theory, science's aspiration to explain the world fully remains intact. The strangeness of superposition is, as Deutsch explains it, simply "the phenomenon of physical variables having different values in different universes." And entanglement, which so bothered Einstein and others, especially for its implication that particles could instantly communicate regardless of their distance in space or time, is also resolved. Information that seemed to travel faster than the speed of light and along no detectable pathway—spookily transmitted as if via ESP—can, in Many Worlds theory, be understood to move differently. Information still spreads through direct contact—the "ordinary" way; it's just that we need to adjust to that contact being via the tangencies of abutting universes. As a further bonus, in Many Worlds theory, randomness goes away, too. A 10 percent chance of an atom decaying is not arbitrary at all, but rather refers to the certainty that the atom will decay in 10 percent of the universes branching from that point. (This being science, there's the glory of nuanced dissent around the precise meaning of each descriptive term, from "chance" to "branching" to "universe.")

In the 1970s, Everett's theory received some of the serious attention it missed at its conception, but today the majority of physicists are not much compelled. "I've never myself subscribed to that view," DiVincenzo says, "but it's not a harmful view." Another quantum-computing physicist called it "completely ridiculous," but Ekert said, "Of all the weird theories out there, I would say Many Worlds is the least weird." In Deutsch's view, "Everett's approach was to look at quantum theory and see what it actually said, rather than hope it said certain things. What we want is for a theory to conform to reality, and in order to find out whether it does, you need to see what the theory actually says. Which with the deepest theories is actually quite difficult, because they violate our intuitions."

I told Deutsch that I'd heard that even Everett thought his theory could never be tested.

"That was a catastrophic mistake," Deutsch said. "Every innovator starts out with the world view of the subject as it was before his innovation. So he can't be blamed for regarding his theory as an interpretation. But"—and here he paused for a moment—"I proposed a test of the Everett theory."

Deutsch posited an artificial-intelligence program run on a computer that could be used in a quantum-mechanics experiment as an "observer"; the AI program, rather than a scientist, would be doing the problematic "looking," and, by means of a clever idea that Deutsch came up with, a physicist looking at the AI observer would see one result if Everett's theory was right and another if the theory was wrong.

It was a thought experiment, though. No AI program existed that was anywhere near sophisticated enough to act as the observer. Deutsch argued that theoretically there could be such a program, though it could only be run on radically more advanced hardware —hardware that could model any other hardware, including that of the human brain. The computer on which the AI program would run "had to have the property of being universal . . . so I had to postulate this quantum-coherent universal computer, and that was really my first proposal for a quantum computer. Though I didn't think of it as that. And I didn't call it a quantum computer. But that's what it was." Deutsch had, it seems, come up with the idea for a quantum computer twice: once in devising a way to test the validity of the Many Worlds Interpretation, and a second time, emerging from the complexity-theory conversation, with evidenced argument supporting Many Worlds as a consequence.

To those who find the Many Worlds Interpretation needlessly baroque, Deutsch writes, "the quantum theory of parallel universes is not the problem—it is the solution. . . . It is the explanation—the only one that is tenable—of a remarkable and counterintuitive reality." The theory also explains how quantum computers might work. Deutsch told me that a quantum computer would be "the first technology that allows useful tasks to be performed in collaboration between parallel universes." The quantum computer's processing power would come from a kind of outsourcing of work, in which calculations literally take place in other universes. Entangled particles would function as paths of communication among different universes, sharing information and gathering the results. So, for example, with the case of Shor's algorithm, Deutsch said, "When we run such an algorithm, countless instances of us are also running it in other universes. The computer then differentiates some of those universes (by creating a superposition), and as a result they perform part of the computation on a huge variety of different inputs. Later those values affect each other, and thereby all contribute to the final answer, in just such a way that the same answer appears in all the universes."

Deutsch is mainly interested in the building of a quantum computer for its implications for fundamental physics, including the Many Worlds Interpretation, which would be a victory for the argument that science can explain the world and that consequently reality is knowable. ("House cures people," Deutsch said to me when discussing Hugh Laurie, "because he's interested in solving problems, not because he's interested in people.") Shor's algorithm excites Deutsch, but here is how his excitement comes through in his book *The Fabric of Reality:*

> To those who still cling to a single-universe world-view, I issue this challenge: *explain how Shor's algorithm works.* I do not merely mean predict that it will work, which is merely a matter of solving a few uncontroversial equations. I mean provide an explanation. When Shor's algorithm has factorized a number, using 10^{500} or so times the computational resources than can be seen to be present, where was the number factorized? There are only about 10^{80} atoms in the entire visible universe, an utterly minuscule number compared with 10^{500}. So if the visible universe were the extent of physical reality, physical reality would not even remotely contain the resources required to factorize such a large number. Who did factorize it, then? How, and where, was the computation performed?

Deutsch believes that quantum computing and Many Worlds are inextricably bound. He is nearly alone in this conviction, though many (especially around Oxford) concede that the construction of a sizable and stable quantum computer might be evidence in favor of the Everett interpretation. "Once there are actual quantum computers," Deutsch said to me, "and a journalist can go to the actual labs and ask how does that actual machine work, the physicists in question will then either talk some obfuscatory nonsense or will explain it in terms of parallel universes. Which will be newsworthy. Many Worlds will then become part of our culture. Really, it has nothing to do with making the computers. But psychologically it has everything to do with making them."

It's tempting to view Deutsch as a visionary in his devotion to the Many Worlds Interpretation, for the simple reason that he has been a visionary before. "Quantum computers should have been invented in the nineteen-thirties," he observed near the end of our conversation. "The stuff that I did in the late nineteen-seventies and early nineteen-eighties didn't use any innovation that hadn't been known in the thirties." That is straightforwardly true. Deutsch went on, "The question is why."

DiVincenzo offered a possible explanation. "Your average physicists will say, 'I'm not strong in philosophy and I don't really know what to think, and it doesn't matter.'" He does not subscribe to Many Worlds but is reluctant to dismiss Deutsch's belief in it, partly because it has led Deutsch to come up with his important theories, but also because "quantum mechanics does have a unique place in physics, in that it does have a subcurrent of philosophy you don't find even in Newton's laws of gravity. But the majority of physicists say it's a quagmire they don't want to get into—they'd rather work out the implications of ideas; they'd rather calculate something."

At Yale, a team led by Robert Schoelkopf has built a two-qubit quantum computer. "Deutsch is an original thinker, and those early papers remain very important," Schoelkopf told me. "But what we're doing here is trying to develop hardware, to see if these descriptions that theorists have come up with work." They have configured their computer to run what is known as a Grover's algorithm, one that deals with a four-card-monte type of question: Which hidden card is the queen? It's a sort of Shor's algorithm for beginners, something that a small quantum computer can take on.

The Yale team fabricates their qubit processor chips in house. "The chip is basically made of a very thin wafer of sapphire or silicon—something that's a good insulator—that we then lay a patterned film of superconducting metal on to form the wiring and qubits," Schoelkopf said. What they showed me was smaller than a pinkie nail and looked like a map of a subway system.

Schoelkopf and his colleague Michel Devoret, who leads a separate team, took me to a large room of black lab benches, inscrutable equipment, and not particularly fancy monitors. The aesthetic was inadvertent steampunk. The dust in the room made me sneeze. "We don't like the janitors to come sweep for fear they'll disturb something," Schoelkopf said.

The qubit chip is small, but its supporting apparatus is imposing. The largest piece of equipment is the plumbing of the very high-end refrigerator, which reduces the temperature around the two qubits to 10 millidegrees above absolute zero. The cold improves the computer's fidelity. Another apparatus produces the microwave signals that manipulate the qubits and set them into any degree of superposition that an experimenter chooses.

Running this Grover's algorithm takes a regular computer three or fewer steps—if, after checking the third card, you still haven't found the queen, you know she is under the fourth card—and on average it takes 2.25 steps. A quantum computer can run it in just one step. This is because the qubits can represent different values at the same time. In the four-card-monte example, each of the cards is represented by one of four states: 0,0; 0,1; 1,0; 1,1. Schoelkopf designates one of these states as the queen, and the quantum computer must determine which one. "The magic comes from the initial state of the computer," he explained. Both of the qubits are set up, via pulses of microwave radiation, in a superposition of zero and one, so that each qubit represents two states at once, and together the two qubits represent all four states.

"Information can, in a way, be holographically represented across the whole computer; that's what we exploit," Devoret explained. "This is a property you don't find in a classical information processor. A bit has to be in one state—it has to be here or there. It's useful to have the bit be everywhere."

Through superposition and entanglement, the computer simultaneously investigates each of the four possible queen locations. "Right now we only get the right answer eighty percent of the time, and we find even that pretty exciting," Schoelkopf said.

With Grover's algorithm, or theoretically with Shor's, calculations are performed in parallel, though not necessarily in parallel worlds. "It's as if I had a gazillion classical computers that were all testing different prime factors at the same time," Schoelkopf summarized. "You start with a well-defined state, and you end with a well-defined state. In between, it's a crazy entangled state, but that's fine."

Schoelkopf emphasized that quantum mechanics is a funny system but that it really is correct. "These oddnesses, like superposition and entanglement—they seemed like limitations, but in fact they are exploitable resources. Quantum mechanics is no longer a new or surprising theory that should strike us as odd."

Schoelkopf seemed to suggest that existential questions like those that Many Worlds poses might be, finally, simply impracticable. "If you have to describe a result in my lab in terms of the computing chip," he continued, "plus the measuring apparatus, plus the computer doing data collection, plus the experimenter at the bench . . . at some point you just have to give up and say, Now quantum mechanics doesn't matter anymore, now I just need a classical result. At some point you have to simplify, you have to throw out some of the quantum information." When I asked him what he thought of Many Worlds and of "collapse" interpretations—in which "looking" provokes a shift from an entangled to an unentangled state—he said, "I have an alternate language which I prefer in describing quantum mechanics, which is that it should really be called Collapse of the Physicist." He knows it's a charming formulation, but he does mean something substantive in saying it. "In reality it's about where to collapse the discussion of the problem."

I thought Deutsch might be excited by the Yale team's research, and I e-mailed him about the progress in building quantum computers. "Oh, I'm sure they'll be useful in all sorts of ways," he replied. "I'm really just a spectator, though, in experimental physics."

Sir Arthur Conan Doyle never liked detective stories that built their drama by deploying clues over time. Conan Doyle wanted to write stories in which all the ingredients for solving the crime were there from the beginning, and in which the drama would be, as in the Poe stories that he cited as precedents, in the mental workings of his ideal ratiocinator. The story of quantum computing follows a Holmesian arc, since all the clues for devising a quantum computer have been there essentially since the discovery of quantum mechanics, waiting for a mind to properly decode them.

But writers of detective stories have not always been able to hew to the rationality of their idealized creations. Conan Doyle believed in "spiritualism" and in fairies, even as the most famed spiritualists and fairy photographers kept revealing themselves to be fakes. Conan Doyle was also convinced that his friend Harry Houdini had supernatural powers; Houdini could do nothing to persuade him otherwise. Conan Doyle just *knew* that there was a spirit world out there, and he spent the last decades of his life corralling evidence ex post facto to support his unshakable belief.

Physicists are ontological detectives. We think of scientists as wholly rational, open to all possible arguments. But to begin with a conviction and then to use one's intellectual prowess to establish support for that conviction is a methodology that really *has* worked for scientists, including Deutsch. One could argue that he dreamed up quantum computing because he was devoted to the idea that science can explain the world. Deutsch would disagree.

In *The Fabric of Reality,* Deutsch writes, "I remember being told, when I was a small child, that in ancient times it was still possible to know *everything that was known*. I was also told that nowadays so much is known that no one could conceivably learn more than a tiny fraction of it, even in a long lifetime. The latter proposition surprised and disappointed me. In fact, I refused to believe it." Deutsch's life's work has been an attempt to support that intuitive disbelief—a gathering of argument for a conviction he held because he just knew.

Deutsch is adept at dodging questions about where he gets his ideas. He joked to me that they came from going to parties, though I had the sense that it had been years since he'd been to one. He said, "I don't like the style of science reporting that goes over that kind of thing. It's misleading. So Brahms lived on black coffee and forced himself to write a certain number of lines of music a day. Look," he went on, "I can't stop you from writing an article about a weird English guy who thinks there are parallel universes. But I think that style of thinking is kind of a putdown to the reader. It's almost like saying, If you're not weird in these ways, you've got no hope as a creative thinker. That's not true. The weirdness is only superficial."

Talking to Deutsch can feel like a case study of reason following desire; the desire is to be a creature of pure reason. As he said in praise of Freud, "He did a good service to the world. He made it OK to speak about the mechanisms of the mind, some of which we may not be aware of. His actual theory was all false, there's hardly a single true thing he said, but that's not so bad. He was a pioneer, one of the first who tried to think about things rationally."

JOSHUA DAVIS

The Crypto-Currency

FROM *The New Yorker*

THERE ARE LOTS of ways to make money: you can earn it, find it, counterfeit it, steal it. Or, if you're Satoshi Nakamoto, a preternaturally talented computer coder, you can invent it. That's what he did on the evening of January 3, 2009, when he pressed a button on his keyboard and created a new currency called bitcoin. It was all bit and no coin. There was no paper, copper, or silver—just 31,000 lines of code and an announcement on the Internet.

Nakamoto, who claimed to be a thirty-six-year-old Japanese man, said he had spent more than a year writing the software, driven in part by anger over the recent financial crisis. He wanted to create a currency that was impervious to unpredictable monetary policies as well as to the predations of bankers and politicians. Nakamoto's invention was controlled entirely by software, which would release a total of 21 million bitcoins, almost all of them over the next twenty years. Every ten minutes or so, coins would be distributed through a process that resembled a lottery. Miners—people seeking the coins—would play the lottery again and again; the fastest computer would win the most money.

Interest in Nakamoto's invention built steadily. More and more people dedicated their computers to the lottery, and forty-four exchanges popped up, allowing anyone with bitcoins to trade them for official currencies like dollars or euros. Creative computer engineers could mine for bitcoins; anyone could buy them. At first a single bitcoin was valued at less than a penny. But merchants gradually began to accept bitcoins, and at the end of 2010 their value began to appreciate rapidly. By June 2011, a bitcoin was worth more than $29. Market gyrations followed, and by September the exchange rate had fallen to $5. Still, with more than 7 million bitcoins in circulation, Nakamoto had created $35 million of value.

And yet Nakamoto himself was a cipher. Before the debut of bitcoin, there was no record of any coder with that name. He used an e-mail address and web site that were untraceable. In 2009 and 2010, he wrote hundreds of posts in flawless English, and though he invited other software developers to help him improve the code, and corresponded with them, he never revealed a personal detail. Then in April 2011, he sent a note to a developer saying that he had "moved on to other things." He has not been heard from since.

When Nakamoto disappeared, hundreds of people posted theories about his identity and whereabouts. Some wanted to know if he could be trusted. Might he have created the currency in order to hoard coins and cash out? "We can effectively think of 'Satoshi Nakamoto' as being on top of a Ponzi scheme," George Ou, a blogger and technology commentator, wrote.

It appeared, though, that Nakamoto was motivated by politics, not crime. He had introduced the currency just a few months after the collapse of the global banking sector, and he published a five-hundred-word essay about traditional fiat, or government-backed, currencies. "The root problem with conventional currency is all the trust that's required to make it work," he wrote. "The central bank must be trusted not to debase the currency, but the history of fiat currencies is full of breaches of that trust. Banks must be trusted to hold our money and transfer it electronically, but they lend it out in waves of credit bubbles with barely a fraction in reserve."

Banks, however, do much more than lend money to overzealous homebuyers. They also, for example, monitor payments so that no one can spend the same dollar twice. Cash is immune to this problem: you can't give two people the same bill. But with digital currency there is the danger that someone can spend the same money any number of times.

Nakamoto solved this problem using innovative cryptography. The bitcoin software encrypts each transaction—the sender and the receiver are identified only by a string of numbers—but a public record of every coin's movement is published across the entire network. Buyers and sellers remain anonymous, but everyone can see that a coin has moved from A to B, and Nakamoto's code can prevent A from spending the coin a second time.

Nakamoto's software would allow people to send money directly to each other without an intermediary, and no outside party could create more bitcoins. Central banks and governments played no role. If Nakamoto ran the world, he would have just fired Ben Bernanke, closed the European Central Bank, and shut down Western Union. "Everything is based on crypto proof instead of trust," Nakamoto wrote in his 2009 essay.

Bitcoin, however, was doomed if the code was unreliable. Earlier this year, Dan Kaminsky, a leading Internet security researcher, investigated the currency and was sure he would find major weaknesses. Kaminsky is famous among hackers for discovering, in 2008, a fundamental flaw in the Internet that would have allowed a skilled coder to take over any web site or even to shut down the Internet. Kaminsky alerted the Department of Homeland Security and executives at Microsoft and Cisco to the problem and worked with them to patch it. He is one of the most adept practitioners of "penetration testing," the art of compromising the security of computer systems at the behest of owners who want to know their vulnerabilities. Bitcoin, he felt, was an easy target.

"When I first looked at the code, I was sure I was going to be able to break it," Kaminsky said, noting that the programming style was dense and inscrutable. "The way the whole thing was formatted was insane. Only the most paranoid, painstaking coder in the world could avoid making mistakes."

Kaminsky lives in Seattle, but while visiting family in San Francisco in July, he retreated to the basement of his mother's house to work on his bitcoin attacks. In a windowless room jammed with computers, Kaminsky paced around talking to himself, trying to build a mental picture of the bitcoin network. He quickly identified nine ways to compromise the system and scoured Nakamoto's code for an insertion point for his first attack. But when he found the right spot, there was a message waiting for him. "Attack Removed," it said. The same thing happened over and over, infuriating Kaminsky. "I came up with beautiful bugs," he said. "But every time I went after the code there was a line that addressed the problem."

He was like a burglar who was certain that he could break into a bank by digging a tunnel, drilling through a wall, or climbing down a vent, and on each attempt he discovered a freshly poured cement barrier with a sign telling him to go home. "I've never seen anything like it," Kaminsky said, still in awe.

Kaminsky ticked off the skills Nakamoto would need to pull it off. "He's a world-class programmer, with a deep understanding of the C++ programming language," he said. "He understands economics, cryptography, and peer-to-peer networking."

"Either there's a team of people who worked on this," Kaminsky said, "or this guy is a genius."

Kaminsky wasn't alone in this assessment. Soon after creating the currency, Nakamoto posted a nine-page technical paper describing how bitcoin would function. That document included three references to the work of Stuart Haber, a researcher at H.P. Labs in Princeton. Haber is a director of the International Association for Cryptologic Research and knew all about bitcoin. "Whoever did this had a deep understanding of cryptography," Haber said when I called. "They've read the academic papers, they have a keen intelligence, and they're combining the concepts in a genuinely new way."

Haber noted that the community of cryptographers is very small: about three hundred people a year attend the most important conference, an annual gathering in Santa Barbara. In all likelihood, Nakamoto belonged to this insular world. If I wanted to find him, the Crypto 2011 conference would be the place to start.

"Here we go, team!" a cheerleader shouted before two burly guys heaved her into the air.

It was a foggy Monday morning in mid-August, and dozens of college cheerleaders had gathered on the athletic fields of the University of California at Santa Barbara for a three-day training camp. Their hollering could be heard on the steps of a nearby lecture hall, where a group of bleary-eyed cryptographers, dressed in shorts and rumpled T-shirts, muttered about symmetric-key ciphers over steaming cups of coffee.

This was Crypto 2011, and the list of attendees included representatives from the National Security Agency, the US military, and an assortment of foreign governments. Cryptographers are little known outside this hermetic community, but our digital safety depends on them. They write the algorithms that conceal bank files, military plans, and your e-mail.

I approached Phillip Rogaway, the conference's program chair. He is a friendly, diminutive man who is a professor of cryptography at the University of California at Davis and who has also taught at Chiang Mai University in Thailand. He bowed when he shook my hand, and I explained that I was trying to learn more about what it would take to create bitcoin. "The people who know how to do that are here," Rogaway said. "It's likely I either know the person or know their work." He offered to introduce me to some of the attendees.

Nakamoto had good reason to hide: people who experiment with currency tend to end up in trouble. In 1998 a Hawaiian resident named Bernard von NotHaus began fabricating silver and gold coins that he dubbed Liberty Dollars. Nine years later, the US government charged NotHaus with "conspiracy against the United States." He was found guilty and is awaiting sentencing. "It is a violation of federal law for individuals . . . to create private coin or currency systems to compete with the official coinage and currency of the United States," the FBI announced at the end of the trial.

Online currencies aren't exempt. In 2007 the federal government filed charges against e-Gold, a company that sold a digital currency redeemable for gold. The government argued that the project enabled money laundering and child pornography, since users did not have to provide thorough identification. The company's owners were found guilty of operating an unlicensed money-transmitting business, and the CEO was sentenced to months of house arrest. The company was effectively shut down.

Nakamoto seemed to be doing the same things as these other currency developers who ran afoul of authorities. He was competing with the dollar, and he ensured the anonymity of users, which made bitcoin attractive for criminals. This winter a web site was launched called Silk Road, which allowed users to buy and sell heroin, LSD, and marijuana as long as they paid in bitcoin.

Still, Lewis Solomon, a professor emeritus at George Washington University Law School, who has written about alternative currencies, argues that creating bitcoin might be legal. "Bitcoin is in a gray area, in part because we don't know whether it should be treated as a currency, a commodity like gold, or possibly even a security," he says.

Gray areas, however, are dangerous, which may be why Nakamoto constructed bitcoin in secret. It may also explain why he built the code with the same peer-to-peer technology that facilitates the exchange of pirated movies and music: users connect with each other instead of with a central server. There is no company in control, no office to raid, and nobody to arrest.

Today bitcoins can be used online to purchase beef jerky and socks made from alpaca wool. Some computer retailers accept them, and you can use them to buy falafel from a restaurant in Hell's Kitchen. In late August, I learned that bitcoins could also get me a room at a Howard Johnson hotel in Fullerton, California, ten minutes from Disneyland. I booked a reservation for my four-year-old daughter and me and received an e-mail from the hotel requesting a payment of 10.305 bitcoins.

By this time, it would have been pointless for me to play the bitcoin lottery, which is set up so that the difficulty of winning increases the more people play it. When bitcoin launched, my laptop would have had a reasonable chance of winning from time to time. Now, however, the computing power dedicated to playing the bitcoin lottery exceeds that of the world's most powerful supercomputer. So I set up an account with Mt. Gox, the leading bitcoin exchange, and transferred $120. A few days later, I bought 10.305 bitcoins with the press of a button and just as easily sent them to the Howard Johnson.

It was a simple transaction that masked a complex calculus. In 1971 Richard Nixon announced that US dollars could no longer be redeemed for gold. Ever since, the value of the dollar has been based on our faith in it. We trust that dollars will be valuable tomorrow, so we accept payment in dollars today. Bitcoin is similar: you have to trust that the system won't get hacked and that Nakamoto won't suddenly emerge to somehow plunder it all. Once you believe in it, the actual cost of a bitcoin—five dollars or thirty?—depends on factors such as how many merchants are using it, how many might use it in the future, and whether or not governments ban it.

My daughter and I arrived at the Howard Johnson on a hot Friday afternoon and were met in the lobby by Jefferson Kim, the hotel's cherubic twenty-eight-year-old general manager. "You're the first person who's ever paid in bitcoin," he said, shaking my hand enthusiastically.

Kim explained that he had started mining bitcoins two months earlier. He liked that the currency was governed by a set of logical rules, rather than the mysterious machinations of the Federal Reserve. A dollar today, he pointed out, buys you what a nickel bought a century ago, largely because so much money has been printed. And, he asked, why trust a currency backed by a government that is $14 trillion in debt?

Kim had also figured that bitcoin mining would be a way to make up the $1,200 he'd spent on a high-performance gaming computer. So far, he'd made only $400, but it was fun to be a pioneer. He wanted bitcoin to succeed, and in order for that to happen, businesses needed to start accepting it.

The truth is that most people don't spend the bitcoins they buy; they hoard them, hoping that they will appreciate. Businesses are afraid to accept them, because they're new and weird—and because the value can fluctuate wildly. (Kim immediately exchanged the bitcoins I sent him for dollars to avoid just that risk.) Still, the currency is young and has several attributes that appeal to merchants. Robert Schwarz, the owner of a computer-repair business in Klamath Falls, Oregon, began selling computers for bitcoin to sidestep steep credit card fees, which he estimates cost him 3 percent on every transaction. "One bank called me saying they had the lowest fees," Schwarz said. "I said, 'No, you don't. Bitcoin does.'" Because bitcoin transfers can't be reversed, merchants also don't have to deal with credit card charge-backs from dissatisfied customers. Like cash, it's gone once you part with it.

At the Howard Johnson, Kim led us to the check-in counter. The lobby featured imitation-crystal chandeliers, ornately framed oil paintings of Venice, and, inexplicably, a pair of faux elephant tusks painted gold. Kim explained that he hadn't told his mother, who owned the place, that her hotel was accepting bitcoins: "It would be too hard to explain what a bitcoin is." He said he had activated the tracking program on his mother's Droid, and she was currently about six miles away. Today, at least, there was no danger of her finding out about her hotel's financial innovation. The receptionist handed me a room card, and Kim shook my hand. "So just enjoy your stay," he said.

Nakamoto's extensive online postings have some distinctive characteristics. First of all, there is the flawless English. Over the course of two years, he dashed off about 80,000 words—the approximate length of a novel—and made only a few typos. He covered topics ranging from the theories of the Austrian economist Ludwig von Mises to the history of commodity markets. Perhaps most interestingly, when he created the first fifty bitcoins, now known as the "genesis block," he permanently embedded a brief line of text into the data: "The Times 03/Jan/2009 Chancellor on brink of second bailout for banks."

This is a reference to a *Times* of London article indicating that the British government had failed to stimulate the economy. Nakamoto appeared to be saying that it was time to try something new. The text, hidden amid a jumble of code, was a sort of digital battle cry. It also indicated that Nakamoto read a British newspaper. He used British spelling ("favour," "colour," "grey," "modernised") and at one point described something as being "bloody hard." An apartment was a "flat," math was "maths," and his comments tended to appear after normal business hours ended in the United Kingdom. In an initial post announcing bitcoin, he employed American-style spelling. But after that a British style appeared to flow naturally.

I had this in mind when I started to attend the lectures at the Crypto 2011 conference, including ones with titles such as "Leftover Hash Lemma, Revisited" and "Time-Lock Puzzles in the Random Oracle Model." In the back of a darkened auditorium, I stared at the attendee list. A Frenchman onstage was talking about testing the security of encryption systems. The most effective method, he said, is to attack the system and see if it fails. I ran my finger past dozens of names and addresses, circling residents of the United Kingdom and Ireland. There were nine.

I soon discovered that six were from the University of Bristol, and they were all together at one of the conference's cocktail parties. They were happy to chat but entirely dismissive of bitcoin, and none had worked with peer-to-peer technology. "It's not at all interesting to us," one of them said. The two other cryptographers from Britain had no history with large software projects. Then I started looking into a man named Michael Clear.

Clear was a young graduate student in cryptography at Trinity College in Dublin. Many of the other research students at Trinity posted profile pictures and phone numbers, but Clear's page just had an e-mail address. A web search turned up three interesting details. In 2008 Clear was named the top computer-science undergraduate at Trinity. The next year, he was hired by Allied Irish Banks to improve its currency-trading software, and he coauthored an academic paper on peer-to-peer technology. The paper employed British spelling. Clear was well versed in economics, cryptography, and peer-to-peer networks.

I e-mailed him, and we agreed to meet the next morning on the steps outside the lecture hall. Shortly after the appointed time, a long-haired, square-jawed young man in a beige sweater walked up to me, looking like an early-Zeppelin Robert Plant. With a pronounced brogue, he introduced himself. "I like to keep a low profile," he said. "I'm curious to know how you found me."

I told him I had read about his work for Allied Irish, as well as his paper on peer-to-peer technology, and was interested because I was researching bitcoin. I said that his work gave him a unique insight into the subject. He was wearing rectangular Armani glasses and squinted so much I couldn't see his eyes.

"My area of focus right now is fully homomorphic encryption," he said. "I haven't been following bitcoin lately."

He responded calmly to my questions. He was twenty-three years old and studied theoretical cryptography by himself in Dublin—there weren't any other cryptographers at Trinity. But he had been programming computers since he was ten, and he could code in a variety of languages, including C++, the language of bitcoin. Given that he was working in the banking industry during tumultuous times, I asked how he felt about the ongoing economic crisis. "It could have been averted," he said flatly.

He didn't want to say whether or not the new currency could prevent future banking crises. "It needs to prove itself," he said. "But it's an intriguing idea."

I told him I had been looking for Nakamoto and thought that he might be here at the Crypto 2011 conference. He said nothing. Finally I asked, "Are you Satoshi?"

He laughed but didn't respond. There was an awkward silence.

"If you'd like, I'd be happy to review the design for you," he offered instead. "I could let you know what I think."

"Sure," I said hesitantly. "Do you need me to send you a link to the code?"

"I think I can find it," he said.

Soon after I met Clear, I traveled to Glasgow, Kentucky, to see what bitcoin mining looked like. As I drove into the town of 14,000, I passed shuttered factories and a central square lined with empty storefronts. On Howdy 106.5, a local radio station, a man tried to sell his bed, his television, and his basset hound—all for $110.

I had come to visit Kevin Groce, a forty-two-year-old bitcoin miner. His uncles had a garbage-hauling business and had let him set up his operation at their facility. The dirt parking lot was jammed with garbage trucks, which reeked in the summer sun.

"I like to call it the new moonshining," Groce said in a smooth Kentucky drawl, as he led me into a darkened room. One wall was lined with four-foot-tall homemade computers with blinking green and red lights. The processors inside were working so hard that their temperature had risen to 170 degrees, and heat radiated into the room. Each system was a jumble of wires and hacked-together parts, with a fan from Walmart duct-taped to the top. Groce had built them three months earlier, for $4,000. Ever since, they had generated a steady flow of bitcoins, which Groce exchanged for dollars, averaging about $1,000 per month so far. He figured his investment was going to pay off.

Groce was wiry, with wisps of gray in his hair, and he split his time between working on his dad's farm, repairing laptops at a local computer store, and mining bitcoin. Groce's father didn't understand Kevin's enthusiasm for the new currency and expected him to take over the farm. "If it's not attached to a cow, my dad doesn't think much of it," Groce said.

Groce was engaged to be married, and he planned to use some of his bitcoin earnings to pay for a wedding in Las Vegas later in the year. He had tried to explain to his fiancée how they could afford it, but she doubted the financial prudence of filling a room with bitcoin-mining rigs. "She gets to cussing every time we talk about it," Groce confided. Still, he was proud of the powerful computing center he had constructed. The machines ran nonstop, and he could control them remotely from his iPhone. The arrangement allowed him to cut tobacco with his father and monitor his bitcoin operation at the same time.

Nakamoto knew that competition for bitcoins would eventually lead people to build these kinds of powerful computing clusters. Rather than let that effort go to waste, he designed software that uses the processing power of the lottery players to confirm and verify transactions. As people like Groce try to win bitcoins, their computers are harnessed to analyze transactions and ensure that no one spends money twice. In other words, Groce's backwoods operation functioned as a kind of bank.

Groce, however, didn't look like a guy Wells Fargo would hire. He liked to stay up late at the garbage-hauling center and thrash through Black Sabbath tunes on his guitar. He gave all his computers pet names, like Topper and the Dazzler, and between guitar solos, he tended to them as if they were prize animals. "I grew up milking cows," Groce said. "Now I'm just milking these things."

A week after the Crypto 2011 conference, I received an e-mail from Clear. He said that he would send me his thoughts on bitcoin in a day. He added, "I also think I can identify Satoshi."

The next morning, Clear sent a lengthy e-mail. "It is apparent that the person(s) behind the Satoshi name accumulated a not insignificant knowledge of applied cryptography," he wrote, adding that the design was "elegant" and required "considerable effort and dedication, and programming proficiency." But Clear also described some of bitcoin's weaknesses. He pointed out that users were expected to download their own encryption software to secure their virtual wallets. Clear felt that the bitcoin software should automatically provide such security. He also worried about the system's ability to grow and the fact that early adopters received an outsized share of bitcoins.

"As far as the identity of the author, it would be unfair to publish an identity when the person or persons has/have taken major steps to remain anonymous," he wrote. "But you may wish to talk to a certain individual who matches the profile of the author on many levels."

He then gave me a name.

*

For a few seconds, all I could hear on the other end of the line was laughter. "I would love to say that I'm Satoshi, because bitcoin is very clever," Vili Lehdonvirta said finally. "But it's not me."

Lehdonvirta is a thirty-one-year-old Finnish researcher at the Helsinki Institute for Information Technology. Clear had discovered that Lehdonvirta used to be a video-game programmer and now studies virtual currencies. Clear suggested that he was a solid fit for Nakamoto.

Lehdonvirta, however, pointed out that he has no background in cryptography and limited C++ programming skills. "You need to be a crypto expert to build something as sophisticated as bitcoin," Lehdonvirta said. "There aren't many of those people, and I'm definitely not one of them."

Still, Lehdonvirta had researched bitcoin and worried about it. "The only people who need cash in large denominations right now are criminals," he said, pointing out that cash is hard to move around and store. Bitcoin removes those obstacles while preserving the anonymity of cash. Lehdonvirta is on the advisory board of Electronic Frontier Finland, an organization that advocates for online privacy, among other things. Nonetheless, he believes that bitcoin takes privacy too far. "Only anarchists want absolute, unbreakable financial privacy," he said. "We need to have a back door so that law enforcement can intercede."

But Lehdonvirta admitted that it's hard to stop new technology, particularly when it has a compelling story. And part of what attracts people to bitcoin, he said, is the mystery of Nakamoto's true identity. "Having a mythical background is an excellent marketing trick," Lehdonvirta said.

A few days later, I spoke with Clear again. "Did you find Satoshi?" he asked cheerfully.

I told him that Lehdonvirta had made a convincing denial and that every other lead I'd been working on had gone nowhere. I then took one more opportunity to question him and to explain all the reasons that I suspected his involvement. Clear responded that his work for Allied Irish Banks was brief and of "no importance." He admitted that he was a good programmer, understood cryptography, and appreciated the bitcoin design. But, he said, economics had never been a particular interest of his. "I'm not Satoshi," Clear said. "But even if I was I wouldn't tell you."

The point, Clear continued, is that Nakamoto's identity shouldn't matter. The system was built so that we don't have to trust an individual, a company, or a government. Anybody can review the code, and the network isn't controlled by any one entity. That's what inspires confidence in the system. Bitcoin, in other words, survives because of what you can see and what you can't. Users are hidden, but transactions are exposed. The code is visible to all, but its origins are mysterious. The currency is both real and elusive—just like its founder.

"You can't kill it," Clear said, with a touch of bravado. "Bitcoin would survive a nuclear attack."

Over the summer, bitcoin actually experienced a sort of nuclear attack. Hackers targeted the burgeoning currency, and though they couldn't break Nakamoto's code, they were able to disrupt the exchanges and destroy web sites that helped users store bitcoins. The number of transactions decreased and the exchange rate plummeted. Commentators predicted the end of bitcoin. In September, however, volume began to increase again, and the price stabilized, at least temporarily.

Meanwhile, in Kentucky, Kevin Groce added two new systems to his bitcoin-mining operation at the garbage depot and planned to build a dozen more. Ricky Wells, his uncle and a coowner of the garbage business, had offered to invest $30,000, even though he didn't understand how bitcoin worked. "I'm just a risk-taking son of a bitch and I know this thing's making money," Wells said. "Plus, these things are so damn hot they'll heat the whole building this winter."

To Groce, bitcoin was an inevitable evolution in money. People use printed money less and less as it is, he said. Consumers need something like bitcoin to take its place. "It's like eight-tracks going to cassettes to CDs and now MP3s," he said.

Even though his friends and most of his relatives questioned his enthusiasm, Groce didn't hide his confidence. He liked to wear a T-shirt he designed that had the words *Bitcoin Millionaire* emblazoned in gold on the chest. He admitted that people made fun of him for it. "My fiancée keeps saying she'd rather I was just a regular old millionaire," he said. "But maybe I will be someday, if these rigs keep working for me."

BRIAN CHRISTIAN

Mind vs. Machine

FROM *The Atlantic*

BRIGHTON, ENGLAND, SEPTEMBER 2009. I wake up in a hotel room 5,000 miles from my home in Seattle. After breakfast, I step out into the salty air and walk the coastline of the country that invented my language, though I find I can't understand a good portion of the signs I pass on my way—LET AGREED, one says, prominently, in large print, and it means nothing to me.

I pause and stare dumbly at the sea for a moment, parsing and reparsing the sign. Normally these kinds of linguistic curiosities and cultural gaps intrigue me; today, though, they are mostly a cause for concern. In two hours, I will sit down at a computer and have a series of five-minute instant-message chats with several strangers. At the other end of these chats will be a psychologist, a linguist, a computer scientist, and the host of a popular British technology show. Together they form a judging panel, evaluating my ability to do one of the strangest things I've ever been asked to do.

I must convince them that I'm human.

Fortunately, I *am* human; unfortunately, it's not clear how much that will help.

The Turing Test

Each year for the past two decades, the artificial-intelligence community has convened for the field's most anticipated and controversial event—a meeting to confer the Loebner Prize on the winner of a competition called the Turing Test. The test is named for the British mathematician Alan Turing, one of the founders of computer science, who in 1950 attempted to answer one of the field's earliest questions: can machines think? That is, would it ever be possible to construct a computer so sophisticated that it could actually be said to be thinking, to be intelligent, to have a mind? And if indeed there was someday such a machine, how would we know?

Instead of debating this question on purely theoretical grounds, Turing proposed an experiment. Several judges each pose questions, via computer terminal, to several pairs of unseen correspondents, one a human "confederate," the other a computer program, and attempt to discern which is which. The dialogue can range from small talk to trivia questions, from celebrity gossip to heavy-duty philosophy—the whole gamut of human conversation. Turing predicted that by the year 2000, computers would be able to fool 30 percent of human judges after five minutes of conversation, and that as a result, one would "be able to speak of machines thinking without expecting to be contradicted."

Turing's prediction has not come to pass; however, at the 2008 contest, the top-scoring computer program missed that mark by just a single vote. When I read the news, I realized instantly that the 2009 test in Brighton could be the decisive one. I'd never attended the event, but I felt I had to go—and not just as a spectator but as part of the human defense. A steely voice had risen up inside me, seemingly out of nowhere: *Not on my watch.* I determined to become a confederate.

The thought of going head-to-head (head-to-motherboard?) against some of the world's top AI programs filled me with a romantic notion that as a confederate I would be *defending the human race,* à la Garry Kasparov's chess match against Deep Blue.

During the competition, each of four judges will type a conversation with one of us for five minutes, then with the other, and then will have ten minutes to reflect and decide which one is the human. Judges will also rank all the contestants—this is used in part as a tiebreaking measure. The computer program receiving the most votes and highest ranking from the judges (regardless of whether it passes the Turing Test by fooling 30 percent of them) is awarded the title of the Most Human Computer. It is this title that the research teams are all gunning for, the one with the cash prize (usually $3,000), the one with which most everyone involved in the contest is principally concerned. But there is also, intriguingly, another title, one given to the *confederate* who is most convincing: the Most Human Human award.

One of the first winners, in 1994, was the journalist and science fiction writer Charles Piatt. How'd he do it? By "being moody, irritable, and obnoxious," as he explained in *Wired* magazine—which strikes me as not only hilarious and bleak but, in some deeper sense, a call to arms: how, in fact, do we be the most human we can be—not only under the constraints of the test but in life?

The Importance of Being Yourself

Since 1991 the Turing Test has been administered at the so-called Loebner Prize competition, an event sponsored by a colorful figure: the former baron of plastic roll-up portable disco dance floors, Hugh Loebner. When asked his motives for orchestrating this annual Turing Test, Loebner cites laziness, of all things: his Utopian future, apparently, is one in which unemployment rates are nearly 100 percent and virtually all of human endeavor and industry is outsourced to intelligent machines.

To learn how to become a confederate, I sought out Loebner himself, who put me in touch with contest organizers, to whom I explained that I'm a nonfiction writer of science and philosophy, fascinated by the Most Human Human award. Soon I was on the confederate roster. I was briefed on the logistics of the competition but not much else. "There's not much more you need to know, really," I was told. "You are human, so just be yourself."

Just be yourself has become, in effect, the confederate motto, but it seems to me like a somewhat naive overconfidence in human instincts—or at worst, like fixing the fight. Many of the AI programs we confederates go up against are the result of decades of work. Then again, so are we. But the AI research teams have huge databases of test runs for their programs, and they've done statistical analysis on these archives: the programs know how to deftly guide the conversation away from their shortcomings and toward their strengths, know which conversational routes lead to deep exchange and which ones fizzle. The average off-the-street confederate's instincts—or judge's, for that matter—aren't likely to be so good. This is a strange and deeply interesting point, amply proved by the perennial demand in our society for dating coaches and public-speaking classes. The transcripts from the 2008 contest show the humans to be such wet blankets that the judges become downright apologetic for failing to provoke better conversation: "I feel sorry for the humans behind the screen, I reckon they must be getting a bit bored talking about the weather," one writes; another offers, meekly, "Sorry for being so banal." Meanwhile a computer appears to be charming the pants off one judge, who in no time at all is gushing *LOLs* and smiley-face emoticons. We can do better.

Thus, my intention from the start was to thoroughly disobey the advice to just show up and be myself—I would spend months preparing to give it everything I had.

Ordinarily this notion wouldn't be odd at all, of course—we train and prepare for tennis competitions, spelling bees, standardized tests, and the like. But given that the Turing Test is meant to evaluate *how human* I am, the implication seems to be that being human (and being oneself) is about more than simply showing up.

The Sentence

To understand why our human sense of self is so bound up with the history of computers, it's important to realize that computers used to *be human*. In the early twentieth century, before a "computer" was one of the digital processing devices that permeate our twenty-first-century lives, it was something else: a job description.

From the mid-eighteenth century onward, computers, many of them women, were on the payrolls of corporations, engineering firms, and universities, performing calculations and numerical analysis, sometimes with the use of a rudimentary calculator. These original, human computers were behind the calculations for everything from the first accurate prediction, in 1757, for the return of Halley's Comet—early proof of Newton's theory of gravity—to the Manhattan Project at Los Alamos, where the physicist Richard Feynman oversaw a group of human computers.

It's amazing to look back at some of the earliest papers on computer science and see the authors attempting to explain what exactly these new contraptions were. Turing's paper, for instance, describes the unheard-of "digital computer" by making analogies to a *human* computer:

> The idea behind digital computers may be explained by saying that these machines are intended to carry out any operations which could be done by a human computer.

Of course, in the decades that followed, we know that the quotation marks migrated, and now it is "digital computer" that is not only the default term, but the *literal* one. In the mid-twentieth century, a piece of cutting-edge mathematical gadgetry was said to be "like a computer." In the twenty-first century, it is the human math whiz who is "like a computer." It's an odd twist: we're *like* the thing that used to be *like* us. We imitate our old imitators, in one of the strange reversals in the long saga of human uniqueness.

Philosophers, psychologists, and scientists have been puzzling over the essential definition of human uniqueness since the beginning of recorded history. The Harvard psychologist Daniel Gilbert says that every psychologist must, at some point in his or her career, write a version of what he calls "The Sentence." Specifically, The Sentence reads like this:

> The human being is the only animal that ______.

The story of humans' sense of self is, you might say, the story of failed, debunked versions of The Sentence. Except now it's not just the animals that we're worried about.

We once thought humans were unique for using language, but this seems less certain each year; we once thought humans were unique for using tools, but this claim also erodes with ongoing animal-behavior research; we once thought humans were unique for being able to do mathematics, and now we can barely imagine being able to do what our calculators can.

We might ask ourselves: Is it appropriate to allow our definition of our own uniqueness to be, in some sense, *reactive* to the advancing front of technology? And why is it that we are so compelled to feel unique in the first place?

"Sometimes it seems," says Douglas Hofstadter, a Pulitzer Prize–winning cognitive scientist, "as though each new step toward AI, rather than producing something which everyone agrees is real intelligence, merely reveals what real intelligence is *not*." While at first this seems a consoling position—one that keeps our unique claim to thought intact—it does bear the uncomfortable appearance of a gradual retreat, like a medieval army withdrawing from the castle to the keep. But the retreat can't continue indefinitely. Consider: if everything that we thought hinged on thinking turns out to not involve it, then . . . what is thinking? It would seem to reduce to either an epiphenomenon—a kind of "exhaust" thrown off by the brain—or, worse, an illusion.

Where is the keep of our *selfhood?*

The story of the twenty-first century will be, in part, the story of the drawing and redrawing of these battle lines, the story of *Homo sapiens* trying to stake a claim on shifting ground, flanked by beast and machine, pinned between meat and math.

Is this retreat a good thing or a bad thing? For instance, does the fact that computers are so good at mathematics in some sense *take away* an arena of human activity, or does it *free* us from having to do a nonhuman activity, liberating us into a more human life? The latter view seems to be more appealing, but less so when we begin to imagine a point in the future when the number of "human activities" left for us to be "liberated" into has grown uncomfortably small. What then?

Alan Turing proposed his test as a way to measure technology's progress, but it just as easily lets us measure our own. The Oxford philosopher John Lucas says, for instance, that if we fail to prevent the machines from passing the Turing Test, it will be "not because machines are so intelligent, but because humans, many of them at least, are so wooden."

Beyond its use as a technological benchmark, the Turing Test is, at bottom, about the act of communication. I see its deepest questions as practical ones: How do we connect meaningfully with each other, as meaningfully as possible, within the limits of language and time? How does empathy work? What is the process by which someone enters into our life and comes to mean something to us? These, to me, are the test's most central questions—the most central questions of being human.

Part of what's fascinating about studying the programs that have done well at the Turing Test is seeing how conversation can work in the total absence of emotional intimacy. A look at the transcripts of Turing Tests past is, frankly, a sobering tour of the various ways in which we demur, dodge the question, lighten the mood, change the subject, distract, burn time: what shouldn't pass for real conversation at the Turing Test probably shouldn't be allowed to pass for real conversation in everyday life either.

How to Talk like a Human

When the Turing Test was first proposed in 1950, it was a hypothetical: technology was nowhere near the point at which a practical test was possible. But as we know, it got there; the first conversational computer program to attract significant notice and attention was Eliza, written in 1964 and 1965 by Joseph Weizenbaum at MIT. Modeled after a Rogerian therapist, Eliza worked on a very simple principle: extract key words from the users' own language and pose their statements back to them. ("I am unhappy." "Do you think coming here will help you not to be unhappy?") If in doubt, it might fall back on some completely generic phrases, like "Please go on." This technique of fitting the users' statements into predefined patterns and responding with a prescribed phrasing of its own—called "template matching"—was Eliza's only capacity.

A look at an Eliza transcript reveals how adeptly such an impoverished set of rules can, in the right context, pass at a glance for understanding:

USER: Men are all alike.
ELIZA: In what way?
USER: They're always bugging us about something or other.
ELIZA: Can you think of a specific example?
USER: Well, my boyfriend made me come here.
ELIZA: Your boyfriend made you come here.
USER: He says I'm depressed much of the time.
ELIZA: I am sorry to hear you are depressed.
USER: It's true. I am unhappy.
ELIZA: Do you think coming here will help you not to be unhappy?
USER: I need some help, that much seems certain.
ELIZA: What would it mean to you if you got some help?
USER: Perhaps I could learn to get along with my mother.
ELIZA: Tell me more about your family.

The results were stunning: many of the people who first talked with Eliza were convinced that they were having a genuine human exchange. In some cases, even Weizenbaum's own insistence to the contrary was of no use. People asked to be left alone to talk "in private," sometimes for hours, and returned with reports of having had a meaningful therapeutic experience. Meanwhile, academics leapt to conclude that Eliza represented "a general solution to the problem of computer understanding of natural language."

One of the strangest twists to the Eliza story, however, was the reaction of the *medical* community, which decided that Weizenbaum had hit upon something both brilliant and useful. The *Journal of Nervous and Mental Disease,* for example, said of Eliza in 1966: "Several hundred patients an hour could be handled by a computer system designed for this purpose. The human therapist, involved in the design and operation of this system, would not be replaced, but would become a much more efficient man." The famed scientist Carl Sagan, in 1975, concurred: "I can imagine the development of a network of computer psychotherapeutic terminals, something like arrays of large telephone booths, in which, for a few dollars a session, we would be able to talk with an attentive, tested, and largely non-directive psychotherapist."

As for Weizenbaum, appalled and horrified, he did something almost unheard-of: an about-face on his entire career. He pulled the plug on the Eliza project, encouraged his own critics, and became one of science's most outspoken opponents of AI research. But the genie was out of the bottle, and there was no going back. The basic "template matching" skeleton and approach of Eliza has been reworked and implemented in some form or another in almost every chat program since, including the contenders at the 2009 Loebner Prize competition. The enthusiasm—as well as the unease—about these programs has only grown.

The Jerk as Genius

The first Loebner Prize competition was held on November 8, 1991, at the Boston Computer Museum. In its first few years, the contest required each program and human confederate to choose a topic, as a means of limiting the conversation. One of the confederates in 1991 was the Shakespeare expert Cynthia Clay, who was famously deemed a computer by three different judges after a conversation about the playwright. The consensus seemed to be: "No one knows that much about Shakespeare." (For this reason, Clay took her misclassifications as a compliment.)

The programmer Joseph Weintraub chose "Whimsical Conversation" for his PC Therapist III and went on to earn the very first Most Human Computer award. As the program practiced it, "whimsical conversation" was not a topic at all but a manner—a crucial distinction. Weintraub was able to play off a looseness in the contest rules, the natural behavior patterns of chatbots, and people's stereotypes about computers and humans, all at once. Strategically, this was brilliant.

By dutifully and scrupulously providing information in response to the questions asked, Clay demonstrated her knowledge and understanding—but sometimes in a factual, encyclopedic way commonly associated with computer systems. Here is a sample of Clay's conversation:

JUDGE: What is your opinion on Shakespeare's plays?
CONFEDERATE: That's pretty general; would you be more specific? Otherwise, I'll just say I like them.

The computer at the "Whimsical Conversation" terminal stood in stark contrast:

JUDGE: It has been a long day! Now I think I'm ready for some fun. The weather isn't very pleasant today. Do you have any jokes that could lift the clouds?
COMPUTER: Everybody talks about the weather but nobody seems to do much about it. Why do you need to tell me now you think you're ready for some fun? Groannnnnn . . . .
JUDGE: What is the definition of whimsical conversation?
COMPUTER: Didn't you say well anyway I would imagine whimsical conversation is like a series of knight's moves in chess-skrt of zig-zag rather than following any nparticular line.? [Weintraub's program, like many that followed it, faked typos. Any unintentional typing errors in the transcripts that follow have been corrected.] Why do you ask? Was that a clever comment, or what? Are such questions much on your mind?
JUDGE: I like the image of knights moving haphazardly across the chess board, does that mean there is no thought to whimsical conversation?
COMPUTER: I could swear you just said how do you define whimsical?? How do yku define whimsical? Were you always so sick sick sick? Why do you need to tell me you like the image of knights moving haphazardly across the chess board?

"I really thought [PC Therapist] was human, because it . . . was weird and funny, in a normal sort of way," said one of the judges, a Harvard graduate student. Weintraub's program, shifting topics wildly and spouting non sequiturs and canned one-liners, came off as zany, a jokester, a much more "human" personality type. At least I used to think so—before I learned how easy this was to mimic.

As Richard Wallace, three-time winner of the Most Human Computer award ('00, '01, and '04), explains: "Experience with [Wallace's chatbot] ALICE indicates that most casual conversation is "state-less," that is, each reply depends only on the current query, without any knowledge of the history of the conversation required to formulate the reply." Many human conversations function in this way, and it behooves AI researchers to determine which types of conversation are stateless—with each remark depending only on the last—and try to create these very sorts of interactions. It's our job as confederates, as humans, to resist them.

One of the classic stateless-conversation types is the kind of zany free-associative riffing that Weintraub's program, PC Therapist III, employed. Another, it turns out, is verbal abuse.

In May 1989, Mark Humphrys, a twenty-one-year-old University College Dublin undergraduate, put online an Eliza-style program he'd written, called MGonz, and left the building for the day. A user (screen name "Someone") at Drake University in Iowa tentatively sent the message "finger" to Humphrys's account—an early-Internet command that acted as a request for basic information about a user. To Someone's surprise, a response came back immediately: "cut this cryptic shit speak in full sentences." This began an argument between Someone and MGonz that lasted almost an hour and a half. (The best part was undoubtedly when Someone said, "you sound like a goddamn robot that repeats everything.")

Returning to the lab the next morning, Humphrys was stunned to find the log and felt a strange, ambivalent emotion. His program might have just shown how to pass the Turing Test, he thought—but the evidence was so profane that he was afraid to publish it.

Humphrys's twist on the Eliza paradigm was to abandon the therapist persona for that of an abusive jerk; when it lacked any clear cue for what to say, MGonz fell back not on therapy clichés like "How does that make you feel?" but on things like "You are obviously an asshole," or "Ah type something interesting or shut up." It's a stroke of genius because, as becomes painfully clear from reading the MGonz transcripts, argument is stateless—that is, unanchored from all context, a kind of Markov chain of riposte, meta-riposte, meta-meta-riposte. Each remark after the first is only about the previous remark. If a program can induce us to sink to this level, of course it can pass the Turing Test.

Once again, the question of what types of human behavior computers can imitate shines light on how we conduct our own, human lives. Verbal abuse is simply less complex than other forms of conversation. In fact, since reading the papers on MGonz and transcripts of its conversations, I find myself much more able to constructively manage heated conversations. Aware of the stateless, knee-jerk character of the terse remark I want to blurt out, I recognize that that remark has far more to do with a reflex reaction to the very last sentence of the conversation than with either the issue at hand or the person I'm talking to. All of a sudden, the absurdity and ridiculousness of this kind of escalation become quantitatively clear, and, contemptuously unwilling to act like a bot, I steer myself toward a more "stateful" response: better living through science.

Beware of Banality

Entering the Brighton Centre, I found my way to the Loebner Prize contest room. I saw rows of seats, where a handful of audience members had already gathered; up front, what could only be the bot programmers worked hurriedly, plugging in tangles of wires and making the last flurries of keystrokes. Before I could get too good a look at them, this year's test organizer, Philip Jackson, greeted me and led me behind a velvet curtain to the confederate area. Out of view of the audience and the judges, the four of us confederates sat around a rectangular table, each at a laptop set up for the test: Doug, a Canadian linguistics researcher; Dave, an American engineer working for Sandia National Laboratories; Olga, a speech-research graduate student from South Africa; and me. As we introduced ourselves, we could hear the judges and audience members slowly filing in, but couldn't see them around the curtain. A man zoomed by in a green floral shirt, talking a mile a minute and devouring finger sandwiches. Though I had never met him before, I knew instantly he could be only one person: Hugh Loebner. Everything was in place, he told us between bites, and the first round of the test would start momentarily. We four confederates grew quiet, staring at the blinking cursors on our laptops. My hands were poised over the keyboard, like a nervous gunfighter's over his holsters.

The cursor, blinking. I, unblinking. Then all at once, letters and words began to materialize:

Hi how are you doing?

The Turing Test had begun.

I had learned from reading past Loebner Prize transcripts that judges come in two types: the small-talkers and the interrogators. The latter go straight in with word problems, spatial-reasoning questions, deliberate misspellings. They lay down a verbal obstacle course, and you have to run it. This type of conversation is extraordinarily hard for programmers to prepare against, because anything goes—and this is why Turing had language and conversation in mind as his test, because they are really a test of everything. The downside to the give-'em-the-third-degree approach is that it doesn't leave much room to express yourself, personality-wise.

The small-talk approach has the advantage of making it easier to get a sense of who a person is—if you are indeed talking to a person. And this style of conversation comes more naturally to layperson judges. For one reason or another, small talk has been explicitly and implicitly encouraged among Loebner Prize judges. It's come to be known as the "strangers on a plane" paradigm. The downside is that these conversations are, in some sense, uniform—familiar in a way that allows a programmer to anticipate a number of the questions.

I started typing back.

CONFEDERATE: hey there!
CONFEDERATE: i'm good, excited to actually be typing
CONFEDERATE: how are you?

I could imagine the whole lackluster conversation spread out before me: *Good. Where are you from? / Seattle. How about yourself? / London.*

Four minutes and 43 seconds left. My fingers tapped and fluttered anxiously.

I could just feel the clock grinding away while we lingered over the pleasantries. I felt this desperate urge to go off script, cut the crap, cut to the chase—because I knew that the computers could do the small-talk thing, which played directly into their preparation. As the generic civilities stretched forebodingly out before me, I realized that this very kind of conversational boilerplate was the enemy, every bit as much as the bots. *How,* I was thinking as I typed another unassuming pleasantry, *do I get an obviously human connection to happen?*

Taking Turns

Part of what I needed to figure out was how to exploit the Loebner Prize's unusual "live typing" medium. The protocol being used was unlike e-mails, text messages, and standard instant-messaging systems in a very crucial way: it transmitted our typing keystroke by keystroke. The judge and I were watching each other type, typos and backspacing and all. I remember some Internet chat programs back in the nineties trying out this character-at-a-time approach, but people for the most part rejected it. It was too invasive, was the feeling: what people like about writing is the time and space to compose and edit a message before sharing it with the other person. The advantage of the character-at-a-time transmission, though, is that it approaches much more closely the condition of speech, with its fluidity of turn-taking and its choppy grammar: what's lost in eloquence is made up for in agility.

It also, then, lets us see typing's "negative space": hesitation. In a chat conversation where text is transmitted with every carriage return, only egregiously long pauses are taken to be part of the interaction. With more fluid and immediate feedback, silence acquires meaning. Failing to quickly answer a question in a face-to-face conversation, for instance, is tantamount in many cases to answering it. I recall asking a friend of mine how things were going with the woman he'd started seeing; the *um* and the tiny pause in which he searched for the right words made it clear they were having problems.

So what did the Loebner Prize's unusual (and recently implemented) protocols enable and disable, compared with the standard, walkie-talkie, turn-taking style? Long missives weren't going to work, as they had in previous years, when programs were able to steamroll the judges by eating up the clock and delivering ridiculously prolix answers. If two parties are taking strict turns under time pressure, they are putting themselves at the mercy of the length of the other's turns. Here, for instance, is one program's conversation transcript from 1996:

> COMPUTER: How about telling me a joke?
> JUDGE: No.
> COMPUTER: Here's my favorite joke . . . A dog limps into a bar with a bloody bandage on his foot. He's wearing a 10-gallon hat, six-shooters in his holsters and chaps. He growled, "I'm a-lookin' for the man that shot my paw."

"The joke's not funny . . ." the judge writes, giving the program an opening to tell another one—which it does ("A knotty, worn-out old string walks into a bar . . ."). Meanwhile, three full minutes have elapsed. If a computer (or confederate) started rambling on too long under the new, live-typing protocols, the judge could and would just cut it off.

And so another piece of my confederate strategy fell into place. I would treat the Turing Test's strange and unfamiliar textual medium more like spoken English and less like the written language. I would attempt to disrupt the turn-taking "wait and parse" pattern that computers understand and create a single, flowing duet of verbal behavior, emphasizing timing. If computers understand little about verbal "harmony," they understand even less about rhythm.

If nothing was happening on my screen, whether or not it was my turn, I'd elaborate a little on my answer, or add a parenthetical, or throw a question back at the judge—just as we offer and/or fill audible silence when we talk out loud. If the judge took too long considering the next question, I'd keep talking. I would be the one (unlike the bots) with something to prove. If I knew what the judge was about to write, I'd spare him the keystrokes and jump in.

There's a trade-off, of course, between the number of opportunities for serve and volley and the sophistication of the responses themselves. The former thrives with brevity, the latter with length. It seemed to me, though, that so much of the nuance (or difficulty) in conversation comes from understanding (or misunderstanding) a question and offering an appropriate (or inappropriate) response—thus, it made sense to maximize the number of interchanges.

Some judges, I discovered, would be startled or confused at this jumping of the gun, and I saw them pause, hesitate, yield, even start backspacing what they had half-written. Other judges cottoned on immediately and leapt right in after me.

In the first round of the 2009 contest, judge Shalom Lappin —a computational linguist at King's College London—spoke with a computer program called Cleverbot and then with me. My strategy of verbosity was clearly in evidence: I made 1,089 keystrokes in five minutes (3.6 keystrokes a second) to Cleverbot's 356 (1.2/sec), and Lappin made 548 keystrokes (1.8/sec) in my conversation, compared with 397 (1.3/sec) with Cleverbot. Not only did I say three times as much as my silicon adversary, but I engaged the judge more, to the tune of 38 percent more typing from Lappin.

Reviewing the logs later, though, I looked for a way to quantify the fluidity of the human-to-human interactions against the rigidity of the human-to-machine ones. It occurred to me that you could count the changes—let's call them "swaps"—of the party who typed the latest keystroke. I did manage to type three times as much as Cleverbot, but the real story, as it turns out, is in the swaps. Lappin's conversation with Cleverbot had 33; his conversation with me had 492, almost fifteen times as many.

The Confederate's Gambit

You don't *converse with* Google, or with most computer systems; you *depose* them. There's a crucial difference. Interestingly, many Loebner Prize judges approach the Turing Test as a kind of cross-examination; strangely, a number of confederates also seem to approach it with that dynamic in mind. One of the human-to-human conversations in 2008 never manages to get out of that stiff question-and-response mode:

JUDGE: Do you have a long drive?
CONFEDERATE: fairly long
JUDGE: so do I :(ah well, do you think you could have used public transport?
CONFEDERATE: i could have
JUDGE: and why not?
CONFEDERATE: i chose not to
JUDGE: that's fair. Do you think we have too many cars on the road or not enough today?
CONFEDERATE: it's not for me to say

Yawn! But the computer in this pair is playful with the judge from the get-go:

> JUDGE: HI
> COMPUTER: Amen to that.
> JUDGE: quite the evangelist
> COMPUTER: Our Father, who art in cyberspace, give us today our daily bandwidth.
> JUDGE: evangelist/nerd lol. So how are things with you today?

And with that, the program has practically sealed up the judge's confidence in its humanity with its second sentence. Note that the confederate's stiff answers prompt more grilling and forced conversation—*what's your opinion on such-and-such political topic?* But with the computer, the judge, misled by the opening wisecracks into assuming it's the real person, remains utterly casual: *how are things?* This makes the contest easier for the computer and harder for the confederate.

It surprised me to see some confederates being coy with their judges. Asked what kind of engineer he is, Dave, to my left, answered, "A good one. :)" And Doug, to my right, responded to a question about what brought him to Brighton with "if I tell you, you'll know immediately that I'm human ;-)" For my money, wit is very successful, but coyness is a double-edged sword. You show a sense of humor, but you jam the cogs of the conversation. Probably the most dangerous thing a confederate can do in a Turing Test is stall. It's suspect—as the guilty party would tend to be the one running out the clock—and it squanders your most precious resource: time.

The humans in a Turing Test are strangers, limited to a medium that is slow and has no vocal tonality, and without much time. A five-second Turing Test would be an easy win for the machines: the judges, barely able to even say "hello," simply wouldn't be able to get enough data from their respondents to make any kind of judgment. A five-hour test would be an easy win for the humans. The Loebner Prize organizers have tried different time limits since the contest's inception, but in recent years they've mostly adhered to Turing's original prescription of five minutes: around the point when conversation starts to get interesting.

A big part of what I needed to do as a confederate was simply to make as much engagement happen in those minutes as I physically and mentally could. Rather than adopt the terseness of a deponent, I offered the prolixity of a writer. In other words, I talked *a lot.* I stopped typing only when to keep going would have seemed blatantly impolite or blatantly suspicious. The rest of the time, my fingers were moving. I went out of my way to embody that maxim of "A bore is a man who, being asked 'How are you?' starts telling you how he is."

JUDGE: Hi, how's things?
CONFEDERATE: hey there
CONFEDERATE: things are good
CONFEDERATE: a lot of waiting, but . . .
CONFEDERATE: good to be back now and going along
CONFEDERATE: how are you?

When we'd finished, and my judge was engaged in conversation with one of my computer counterparts, I strolled around the table, seeing what my comrades were up to. Looking over at my fellow confederate Dave's screen, I noticed his conversation began as if he were on the receiving end of an interrogation, and he was answering in a kind of minimal staccato:

JUDGE: Are you from Brighton?
CONFEDERATE: No, from the US
JUDGE: What are you doing in Brighton?
CONFEDERATE: On business
JUDGE: How did you get involved with the competition?
CONFEDERATE: I answered an e-mail.

Like a good deponent, he let the questioner do all the work. When I saw how stiff Dave was being, I confess I felt a certain confidence—I, in my role as the world's worst deponent, was perhaps in fairly good shape as far as the Most Human Human award was concerned.

This confidence lasted approximately sixty seconds, or enough time for me to continue around the table and see what another fellow confederate, Doug, and his judge had been saying.

JUDGE: Hey Bro, I'm from TO.
CONFEDERATE: cool
CONFEDERATE: leafs suck

CONFEDERATE: ;-)
JUDGE: I am just back from a sabbatical in the CS Dept. at U of T.
CONFEDERATE: nice!
JUDGE: I remember when they were a great team.
JUDGE: That carbon date me, eh?
CONFEDERATE: well, the habs were a great team once, too . . .
CONFEDERATE: *sigh*
JUDGE: YEH, THEY SUCK TOO.
CONFEDERATE: (I'm from Montreal, if you didn't guess)

Doug and his judge had just discovered that they were both Canadian. They let rip with abbreviations and nicknames and slang and local references. And then they started to talk about hockey.

I was in trouble.

Six months after the 2009 contest, a video appeared on YouTube of a man having a shockingly cogent conversation with a bot about Shakespeare's *Hamlet.* Some suspected it might herald a new age for chatbots, and for AI. Others, including myself, were unimpressed. Sophisticated behavior doesn't necessarily indicate a *mind.* It might indicate just a *memory.* As Dali so famously put it, "The first man to compare the cheeks of a young woman to a rose was obviously a poet; the first to repeat it was possibly an idiot."

For instance, Richard Wallace, the three-time Most Human Computer winner, recounts an "AI urban legend" in which "a famous natural language researcher was embarrassed . . . when it became apparent to his audience of Texas bankers that the robot was consistently responding to the *next* question he was about to ask . . . [His] demonstration of natural language understanding . . . was in reality nothing but a simple script."

The moral of the story: no demonstration is ever sufficient. Only *interaction* will do. In the 1997 contest, one judge gets taken for a ride by Catherine, waxing political and really engaging in the topical conversation "she" has been programmed to lead about the Clintons and Whitewater. In fact, everything is going swimmingly until the very end, when the judge signs off:

JUDGE: it looks like i've got to go, Catherine, its been nice talking with you
JUDGE: do you need water or something before the 5 minute rounds?

At which point the program erupts into gibberish.

> COMPUTER: OK, yes on balance . . . Time to get off this one I think and onto something more down to earth!
> JUDGE: like what?
> COMPUTER: Almost certainly! let's move on
> COMPUTER: I suppose it depends on where you're coming from, but as the song goes in My Fair Lady, "Why can't a woman be more like a man?"

Defeat from the jaws of victory.

We so often think of intelligence, of AI, in terms of sophistication, or complexity of behavior. But in so many cases, it's impossible to say much with certainty about the program itself, because any number of different pieces of software—of wildly varying levels of "intelligence"—could have produced that behavior.

No, I think sophistication, complexity of behavior, is not it at all. For instance, you can't judge the intelligence of an orator by the eloquence of his prepared remarks; you must wait until the Q&A and see how he fields questions. The computation theorist Hava Siegelmann once described intelligence as "a kind of sensitivity to things." These Turing Test programs that hold forth may produce interesting output, but they're rigid and inflexible. They are, in other words, insensitive—occasionally fascinating talkers that cannot listen.

As computing technology in the twenty-first century moves increasingly toward mobile devices, we've seen the 1990s' explosive growth in processor speed taper off and product development become less about raw computing horsepower than about the overall design of the product and its fluidity, reactivity, and ease of use. This fascinating shift in computing emphasis may be the cause, effect, or correlative of a healthier view of human intelligence—an understanding, not so much that it is complex and powerful, per se, as that it is reactive, responsive, sensitive, nimble. Our computers, flawed mirrors that they are, have helped us see that about ourselves.

The Most Human Human

The Most Human Computer award in 2009 goes to David Levy and his program, Do-Much-More. Levy, who also won in '97, with

Catherine, is an intriguing guy: he was one of the big early figures in the digital-chess scene of the '70s and '80s, and was one of the organizers of the Marion Tinsley–Chinook checkers matches that preceded the Kasparov–Deep Blue showdowns in the '90s. He's also the author of the recent nonfiction book *Love and Sex with Robots,* to give you an idea of the sorts of things that are on his mind when he's not competing for the Loebner Prize.

Levy stands up, to applause, accepts the award from Philip Jackson and Hugh Loebner, and makes a short speech about the importance of AI for a bright future and the importance of the Loebner Prize for AI. I know what's next on the agenda, and my stomach knots. I'm certain that Doug's gotten it; he and the judge were talking Canada thirty seconds into their conversation.

Ridiculous Canadians and their ice hockey, I'm thinking. Then I'm thinking how ridiculous it is that I'm even allowing myself to get this worked up about some silly award. Then I'm thinking how ridiculous it is to fly 5,000 miles just to have a few minutes' worth of IM conversations. Then I'm thinking how maybe it'll be great to be the runner-up; I can compete again in 2010 in Los Angeles, with the home-field cultural advantage, and finally prove—

"And the results here show also the identification of the humans," Jackson announces, "and from the ranking list we can see that 'Confederate 1,' which is Brian Christian, was the most human."

And he hands me the certificate for the Most Human Human award.

I didn't know how to feel, exactly. It seemed strange to treat the award as meaningless or trivial, but did winning really represent something about me as a person? More than anything, I felt that together, my fellow confederates and I had avenged the mistakes of 2008 in dramatic fashion. That year, the twelve judges decided five times that computer programs were more human than confederates. In three of those instances, the judge was fooled by a program named Elbot, which was the handiwork of a company called Artificial Solutions, one of many new businesses leveraging chatbot technology. One more deception, and Elbot would have tricked 33 percent of that year's dozen judges—surpassing Turing's 30 percent mark and making history. After Elbot's victory at the Loebner Prize and the publicity that followed, the company seemingly decided to prioritize the Elbot software's more commercial applications; at any rate, it had not entered the '09 contest as the returning champion.

In some ways a closer fight would have been more dramatic. Between us, we confederates hadn't permitted a single vote to go the machines' way. Whereas 2008 was a nail biter, 2009 was a rout. We think of science as an unhaltable, indefatigable advance. But in the context of the Turing Test, humans—dynamic as ever—don't allow for that kind of narrative. We don't provide the kind of benchmark that sits still.

As for the prospects of AI, some people imagine the future of computing as a kind of heaven. Rallying behind an idea called the Singularity, people like Ray Kurzweil (in *The Singularity Is Near*) and his cohort of believers envision a moment when we make smarter-than-us machines, which make machines smarter than themselves, and so on, and the whole thing accelerates exponentially toward a massive ultra-intelligence that we can barely fathom. Such a time will become, in their view, a kind of a techno-Rapture, in which humans can upload their consciousness onto the Internet and get assumed—if not bodily, than at least mentally—into an eternal, imperishable afterlife in the world of electricity.

Others imagine the future of computing as a kind of hell. Machines black out the sun, level our cities, seal us in hyperbaric chambers, and siphon our body heat forever.

I'm no futurist, but I suppose, if anything, I prefer to think of the long-term future of AI as a kind of purgatory: a place where the flawed but good-hearted go to be purified—and tested—and come out better on the other side.

Who would have imagined that the computer's earliest achievements would be in the domain of logical analysis, a capacity once held to be what made us most different from everything else on the planet? That it could fly a plane and guide a missile before it could ride a bike? That it could create plausible preludes in the style of Bach before it could make plausible small talk? That it could translate before it could paraphrase? That it could spin half-discernible essays on postmodern theory before it could be shown a chair and say, as most toddlers can, "chair"?

As computers have mastered rarefied domains once thought to be uniquely human, they simultaneously have failed to master the ground-floor basics of the human experience—spatial orientation, object recognition, natural language, adaptive goal-setting—and in so doing, have shown us how impressive, computationally and otherwise, such minute-to-minute fundamentals truly are.

We forget how impressive we are. Computers are reminding us.

One of my best friends was a barista in high school. Over the course of a day, she would make countless subtle adjustments to the espresso being made, to account for everything from the freshness of the beans to the temperature of the machine to the barometric pressure's effect on the steam volume, meanwhile manipulating the machine with an octopus's dexterity and bantering with all manner of customers on whatever topics came up. Then she went to college and landed her first "real" job: rigidly procedural data entry. She thought back longingly to her barista days—when her job actually made demands of her intelligence.

Perhaps the fetishization of analytical thinking, and the concomitant denigration of the creatural—that is, animal—and bodily aspects of life are two things we'd do well to leave behind. Perhaps at last, in the beginnings of an age of AI, we are starting to *center* ourselves again, after generations of living slightly to one side—the logical, left-hemisphere side. Add to this that humans' contempt for "soulless" animals, our unwillingness to think of ourselves as descended from our fellow "beasts," is now challenged on all fronts: growing secularism and empiricism, growing appreciation for the cognitive and behavioral abilities of organisms other than ourselves, and, not coincidentally, the entrance onto the scene of an entity with considerably less soul than we sense in a common chimpanzee or bonobo—in this way AI may even turn out to be a boon for animal rights.

Indeed, it's entirely possible that we've seen the high-water mark of our left-hemisphere bias. I think the return of a more balanced view of the brain and mind—and of human identity—is a good thing, one that brings with it a changing perspective on the sophistication of various tasks.

It's my belief that only experiencing and understanding *truly* disembodied cognition—only seeing the coldness and deadness and disconnectedness of something that really *does* deal in pure abstraction, divorced from sensory reality—can snap us out of it. Only this can bring us, quite literally, back to our senses.

In a 2006 article about the Turing Test, the Loebner Prize cofounder Robert Epstein writes, "One thing is certain: whereas the confederates in the competition will never get any smarter, the computers will." I agree with the latter and couldn't disagree more strongly with the former.

When the world-champion chess player Garry Kasparov defeated Deep Blue, rather convincingly, in their first encounter in 1996, he and IBM readily agreed to return the next year for a rematch. When Deep Blue beat Kasparov (rather less convincingly) in '97, Kasparov proposed another rematch for '98, but IBM would have none of it. The company dismantled Deep Blue, which never played chess again.

The apparent implication is that—because technological evolution seems to occur so much faster than biological evolution (measured in years rather than millennia)—once the *Homo sapiens* species is overtaken, it won't be able to catch up. Simply put: the Turing Test, once passed, is passed forever. I don't buy it.

Rather, IBM's odd anxiousness to get out of Dodge after the '97 match suggests a kind of insecurity on its part that I think proves my point. The fact is, the human race got to where it is by being the most adaptive, flexible, innovative, and quick-learning species on the planet. We're not going to take defeat lying down.

No, I think that while the first year that computers pass the Turing Test will certainly be a historic one, it will not mark the end of the story. Indeed, the *next* year's Turing Test will truly be the one to watch—the one where we humans, knocked to the canvas, must pull ourselves up; the one where we learn how to be better friends, artists, teachers, parents, lovers; the one where we come back. More human than ever.

Contributors' Notes

Other Notable Science and Nature Writing of 2011

Contributors' Notes

Michael Behar is a freelance journalist based in Boulder, Colorado, where he lives with his wife and two-year-old son. He writes about adventure travel, the environment, and innovations in science and technology for more than twenty-five national publications, including *Outside, Wired, Skiing, Men's Journal, Popular Science,* and *OnEarth,* where he is a contributing editor. He is the former science editor for *National Geographic* and served as a senior editor at *Wired* from 1995 to 2000. In addition to this anthology, he's been featured in *The Best American Travel Writing 2006* and *The Best of Technology Writing 2008*.

Deborah Blum is a Pulitzer Prize–winning science journalist, a blogger for *Wired,* and the author of five books, most recently *The Poisoner's Handbook: Murder and the Birth of Forensic Medicine in Jazz Age New York,* a *New York Times* paperback bestseller. She writes for publications ranging from *Scientific American* to the literary journal *Tin House* and is a founding editor of the acclaimed e-book review site Download the Universe. She teaches journalism at the University of Wisconsin–Madison.

Brendan Buhler writes about galaxies, genetics, parasites, clean energy, the skull-strewn wilderness, people who keep leopards as pets, and jumbo Elvis impersonators. Politics, too, but cockroaches and whimsy preferred. His work has appeared in the *Los Angeles Times* and *California* and *Sierra* magazines. The chapter he contributed to *101 Places Not to See Before You Die* was about the annual porn convention in Vegas, an experience that will tarnish your very soul with the orange cheese dust of vileness.

Brian Christian is the author of *The Most Human Human,* a *Wall Street Journal* bestseller and a *New Yorker* favorite book of 2011. His writing has appeared in *The Atlantic, Wired,* the *Wall Street Journal,* the *Guardian,* and the *Paris Review,* as well as scientific journals such as *Cognitive Science* and literary journals such as *AGNI, Gulf Coast,* and *Best New Poets.* He has been featured on *The Charlie Rose Show, The Daily Show,* and NPR's *Radiolab* and has lectured at Google, Microsoft, and the Santa Fe Institute. Christian holds degrees in philosophy, computer science, and poetry from Brown University and the University of Washington.

Jason Daley is a freelance writer specializing in natural history, science, the environment, and travel. His work has appeared in *Discover, Wired, Popular Science, Outside,* and other magazines.

Joshua Davis (www.joshuadavis.net) is a contributing editor at *Wired.* His book *The Underdog* chronicles his experiences as the lightest man to ever compete at the US Sumo Open.

David Dobbs writes for *The Atlantic,* the *New York Times Magazine, National Geographic, Nature,* and other publications. He is currently writing a book with the working title *The Orchid and the Dandelion,* which explores the notion that the genes and traits underlying some of our most torturous mood and behavior problems may also generate some of our greatest strengths and contentment. He is the author of four previous books, most recently the number-one best-selling Kindle Single "My Mother's Lover," which unearths a secret affair his mother had with a doomed flight surgeon in World War II. He blogs on culture, science, and literature at Neuron Culture.

David Eagleman is a neuroscientist at Baylor College of Medicine, where he directs the Laboratory for Perception and Action (EaglemanLab.net) and the Initiative on Neuroscience and Law (NeuLaw.org). He is best known for his work on time perception, vision, synesthesia, social neuroscience, and neurolaw. He is a Guggenheim Fellow, a council member in the World Economic Forum, and a *New York Times* best-selling author, published in twenty-seven languages.

Rivka Galchen is the author of the novel *Atmospheric Disturbances,* which won the William Saroyan International Prize for Writing. Her stories and essays have appeared in such publications as *Harper's Magazine, The Believer, The New Yorker,* and the *New York Times,* and she was recently named to *The New Yorker*'s list of Twenty Writers Under Forty.

Thomas Goetz is the executive editor of *Wired* and the author of *The Decision Tree: A Manifesto for Personalized Medicine.* He holds a master's degree in public health and writes often about the confluence of technology and health care. In particular, his work examines how scalable technologies can help disseminate powerful ideas that improve and save lives. His next book, *The Remedy,* will explore the emergence of the germ theory of disease and the unexpected role of Sherlock Holmes in the attempt to cure tuberculosis. It will be published in 2013.

Jerome Groopman holds the Dina and Raphael Recanati Chair of Medicine at Harvard Medical School and is the chief of experimental medicine at Beth Israel Deaconess Medical Center. He received his BA from Columbia College summa cum laude and his MD from Columbia College of Physicians and Surgeons in New York. He served his internship and residency in internal medicine at Massachusetts General Hospital and his specialty fellowships in hematology and oncology at the University of California, Los Angeles, and Children's Hospital/Sidney Farber Cancer Center, Harvard Medical School, in Boston. He serves on many scientific editorial boards and has published more than 150 scientific articles. In 2000 he was elected to the Institute of Medicine of the National Academies. He writes regularly about biology and medicine for lay audiences as a staff writer at *The New Yorker.* In 2011 he coauthored, with Dr. Pamela Hartzband, *Your Medical Mind: How to Decide What Is Right for You.*

Thomas Hayden has been an oceanographer, a staff writer at *Newsweek* and *US News & World Report,* and a freelance writer for many publications, from *National Geographic* to *Manure Matters.* He teaches science and environmental writing in Stanford University's School of Earth Sciences and Graduate School in Journalism. He is coeditor of the forthcoming *Science Writer's Handbook* and blogs about science with friends at The Last Word on Nothing (www.lastwordonnothing.com).

Virginia Hughes is a freelance writer based in Brooklyn, New York. She writes for a variety of publications, including *Nature, New Scientist, Popular Science, Scientific American,* and the delightfully quirky science blog The Last Word on Nothing (www.lastwordonnothing.com).

David Kirby, a regular contributor to the *Huffington Post* since 2005, has been a professional journalist for twenty years and was also a contract writer for *The New York Times,* where he covered health and science, among many other topics. He has written for several national magazines and was a correspondent in Mexico and Central America from 1986 to 1990. Kirby worked in New York City politics and medical research, including at the American Foundation for AIDS Research (AmFAR). His third book, *Death at SeaWorld: Shamu and the Dark Side of Killer Whales in Captivity,* was published in July 2012. He lives in Brooklyn, New York.

Elizabeth Kolbert is a staff writer for *The New Yorker* and the author of *Field Notes from a Catastrophe: Man, Nature, and Climate Change.* Her series on global warming, "The Climate of Man," from which the book was adapted, won the American Association for the Advancement of Science's magazine-writing award and a National Academies communications award. She is a two-time National Magazine Award winner and has received a Heinz Award and a Lannan Literary Fellowship. Kolbert lives in Williamstown, Massachusetts.

Robert Kunzig is the senior environment editor for *National Geographic.* His article on cities was the last in the magazine's yearlong series on global population trends. A science journalist for thirty years, including fourteen at *Discover,* Kunzig has written two books, *Fixing Climate* (with Wallace Broecker) and *Mapping the Deep,* about oceanography, which won the Aventis/Royal Society prize in 2000. He lives in Birmingham, Alabama, with his wife, Karen Fitzpatrick.

Mark McClusky, the special projects editor for *Wired,* leads the magazine's editorial efforts on new platforms, including its iPad and other tablet editions. From 2005 to 2010, McClusky was the magazine's senior editor for products, directing all coverage of gear and gadgets. He also served as editor of the newsstand-only *Wired Test* magazines and was the founding editor of *Playbook, Wired*'s sports technology blog. McClusky, a coauthor of *Alinea* with the Michelin three-star chef Grant Achatz, is an avid cook who has studied at the Culinary Institute of America in Napa. He lives in Oakland, California, with his wife and two daughters.

Mark W. Moffett is the author of three books, most recently *Adventures Among Ants,* in which he investigates the organization of ant societies. He received his doctorate from Harvard under the environmentalist E. O. Wilson and is currently a research associate in entomology at the National Museum of Natural History (Smithsonian). Moffett received the Explorers Club's Lowell Thomas Medal, a Distinguished Explorer Award from the Roy Chapman Andrews Society, Yale's Poynter Fellowship for Journalism, and Harvard's Bowdoin Prize for writing. Among his many adventures in more than one hundred countries, he has climbed the world's tallest tree, descended into sinkholes a quarter-mile deep to find new frog species, and placed a scorpion on Conan O'Brien's face. Currently he is studying issues of identity in the formation of animal societies.

Sy Montgomery has been chased by an angry silverback gorilla in Zaire and bitten by a vampire bat in Costa Rica, worked in a pit crawling with

18,000 snakes in Manitoba and handled a wild tarantula in French Guiana. She's been deftly undressed by an orangutan in Borneo and swum with pink dolphins, electric eels, and piranhas in the Amazon. She is now at work on *Soul of the Octopus,* her nineteenth book.

Michael Roberts is the senior executive editor at *Outside,* where he has been on staff since 2000. In addition to editing and writing feature stories, he manages the brand's iPad app, Outside+ Magazine. Before turning to journalism, Roberts was employed as a naturalist in the Golden Gate National Recreation Area and as a field biologist in Ecuador, Costa Rica, Alaska, and Baja. He now lives, works, and plays in Marin County, California, with his wife and two sons.

John Seabrook is a staff writer at *The New Yorker.* He is the author of three books, including *Nobrow: The Culture of Marketing, the Marketing of Culture* and *Flash of Genius and Other True Stories of Invention.* He lives in Brooklyn with his wife and two children.

Michael Specter has been a staff writer at *The New Yorker* since 1998. He writes often about science, technology, and global public health. Since joining the magazine, he has written several articles about the global AIDS epidemic, as well as articles about avian influenza, malaria, the world's diminishing freshwater resources, synthetic biology, the attempt to create edible meat in a lab, and the debate over the meaning of our carbon footprint. He has also published profiles of many subjects, including Lance Armstrong, the ethicist Peter Singer, Sean (P. Diddy) Combs, Manolo Blahnik, and Miuccia Prada. Before moving to *The New Yorker,* Specter had been a roving foreign correspondent for the *New York Times,* based in Rome. From 1995 to 1998, Specter served as the *Times* Moscow bureau chief. Before that he worked at the *Washington Post,* where he covered local news from 1985 to 1991, then became the national science reporter and, later, the newspaper's New York bureau chief. Specter has twice received the Global Health Council's annual Excellence in Media Award, for his 2001 article about AIDS, "India's Plague," and for his 2004 article "The Devastation," about the ethics of testing HIV vaccines in Africa. He also received the 2002 AAAS Science Journalism Award for his 2001 article "Rethinking the Brain," about the scientific basis of how we learn. His most recent book, *Denialism: How Irrational Thinking Hinders Scientific Progress, Harms the Planet, and Threatens Our Lives,* received the 2009 Robert P. Balles Annual Prize in Critical Thinking, presented by the Committee for Skeptical Inquiry. In 2011 Specter won the World Health Organization's Stop TB Partnership Award for Excellence in Reporting for his *New Yorker* article "A Deadly Misdiagnosis," about the dangers of inaccurate TB tests in India, which has the highest rate of TB in the world. Specter lives in New York City.

Bijal P. Trivedi is a freelance writer whose work has taken her from the hidden vaults of the Metropolitan Museum of Art to the Serengeti Plains to Moscow's Star City, where she joined space tourism entrepreneurs on the "Vomit Comet" for astronaut training. Her work has appeared in more than twenty-five publications, including *New Scientist, Wired, Discover, Scientific American, The Economist, National Geographic,* and *Air & Space.* Her story "Slimming for Slackers" won the Wistar Institute Science Journalism Award. "Life on Hold," also written for *New Scientist,* won the 2005–2006 Michael E. DeBakey Journalism Award. "The Rembrandt Code," published in *Wired,* was tagged "outstanding story on any subject: print" by the South Asian Journalists Association. Trivedi has taught in New York University's Science, Health, and Environmental Reporting Program. She was born in the UK, raised in Australia, and now lives with her husband, Chad Cohen, and her two wonderful kids in Washington, DC.

Carl Zimmer is the author of twelve books, including *Parasite Rex* and *Soul Made Flesh,* and coauthor, with the biologist Doug Emlen, of the textbook *Evolution: Making Sense of Life.* He writes frequently for the *New York Times, National Geographic, Slate,* and *Discover,* where he is a contributing editor. He has earned numerous awards, including prizes from the American Association for the Advancement of Science and the National Academy of Sciences, for his work. He is a lecturer at Yale University and lives in Guilford, Connecticut, with his wife, Grace Zimmer, and their two daughters.

Other Notable Science and Nature Writing of 2011

SELECTED BY TIM FOLGER

NATALIE ANGIER
Path of Discovery, Lighted by a Bug Man's Insights. *New York Times,* April 4.
So Much More Than Plasma and Poison. *New York Times,* June 6.

NICHOLAS BAKALAR
Study of Vision Tackles a Philosophy Riddle. *New York Times,* April 25.
MARCIA BARTUSIAK
Chemistry Lesson. *Natural History,* November.
Ye Olde Black Hole. *Natural History,* September.
ELIF BATUMAN
Natural Histories. *The New Yorker,* October 24.
BRUCE BARCOTT
Arctic Fever. *OnEarth,* Spring.
Totally Psyched for the Full-Rip Nine. *Outside,* August.
LAURA BEIL
The Magic Bullet for Prostate Cancer. *Men's Health,* March.
YUDHIJIT BHATTACHARJEE
The Fall and Rise of Douglas Prasher. *Discover,* April.
Whales' Grandeur and Grace, Up Close. *New York Times,* April 18.
GEORGE BLACK
Coal on a Roll. *OnEarth,* Fall.
Life and Death in a Dry Land. *OnEarth,* Winter.
BRENDAN BORRELL
The Gloucester Fish War. *Bloomberg Businessweek,* November 22.
The Great Pumpkin. *Smithsonian,* October.

JAMES CAMPBELL
Star Power. *Outside,* December.

KENNETH CHANG
With "Coolest Job Ever" Ending, Astronauts Seek Next Frontier. *New York Times,* April 23.
ADRIAN CHO
Particle Physicists' New Extreme Teams. *Science,* September 16.
TOM CLYNES
Scientist in a Strange Land. *Popular Science,* October.
RICHARD CONNIFF
Dying for Discovery. *New York Times,* January 16.
JENNIFER COUZIN-FRANKEL
Aging Genes: The Sirtuin Story Unfolds. *Science,* December 2.

BRIAN DOYLE
The Creature Beyond the Mountains. *Orion,* September–October.
GARETH DYKE
The Dinosaur Baron of Transylvania. *Scientific American,* October.

TIMOTHY EGAN
Sometimes the Bear Gets You. *New York Times,* September 29.
MARTIN ENSERINK
Can This DNA Sleuth Help Catch Criminals? *Science,* February 18.
HELEN EPSTEIN
Flu Warning: Beware the Drug Companies! *New York Review of Books,* May 12.

MICHAEL FINKEL
The Volcano Next Door. *National Geographic,* April.
DOUGLAS FOX
Omens from a Vanished Sea. *High Country News,* October 31.
Super Storms. *Popular Mechanics,* October.
ERIC FREEDMAN
Extinction Is Forever. *Earth Island,* Autumn.
HOWARD W. FRENCH
E. O. Wilson's Theory of Everything. *The Atlantic,* November.

TED GENOWAYS
Tar Sands Showdown in the Nebraska Sandhills. www.onearth.org, September 30.
DAVID GESSNER
Spoiling Walden: Or How I Learned to Stop Worrying and Love Cape Wind. www.onearth.org, November 29.
JUSTIN GILLIS
A Warming Planet Struggles to Feed Itself. *New York Times,* June 4.
ANNE GISLESON
Shifting: Cycles of Loss on a Sinking Coast. *Ecotone,* Fall.
DAVID GRIMM
Are Dolphins Too Smart for Captivity? *Science,* April 29.

STEPHEN S. HALL
Iceman Unfrozen. *National Geographic,* November.
SUE HALPERN
Over the High-Tech Rainbow. *New York Review of Books,* November 24.
ADAM HIGGINBOTHAM
Half-Life. *Wired,* May.
JOSEPH HOOPER
The Man Who Would Stop Time. *Popular Science,* August.

ROBERT IRION
Something New Under the Sun. *Smithsonian,* April.

FREDERICK KAUFMAN
The Second Green Revolution. *Popular Science,* February.
MARC KAUFMAN
Space Odyssey: Scientists Go to the Extremes of the Earth to Divine the Secrets of Extraterrestrial Life. *Washington Post,* February 27.
JONATHON KEATS
The Search for a More Perfect Kilogram. *Wired,* October.
JEFFREY KLUGER
Little Women. *Time,* October 31.
DAN KOEPPEL
Let There Be LED. *Wired,* September.
ELIZABETH KOLBERT
The Acid Sea. *National Geographic,* April.
KAI KUPFERSCHMIDT
Sharp Insights and a Sharp Tongue. *Science,* November 4.

JOHAN LEHRER
The Truth Wears Off. *The New Yorker,* December 13.
ALAN LIGHTMAN
The Accidental Universe. *Harper's Magazine,* December.

J. B. MACKINNON
Wisdom in the Wild. *Orion,* May–June.
MATTHEW MCGOUGH
The Lazarus File. *The Atlantic,* June.
MAUREEN NANDINI MITRA
Ready or Not. *Earth Island Journal,* Autumn.

MICHELLE NIJHUIS
Crisis in the Caves. *Smithsonian,* July–August.

EMILENE OSTLIND
The Perilous Journey of Wyoming's Migrating Pronghorn. *High Country News,* December 26.

MIKE PEED
We Have No Bananas. *The New Yorker,* January 10.

MICHAEL S. REIDY
The Rucksack of Joseph Dalton Hooker. *Alpinist,* Winter.
NATHANIEL RICH
The Luckiest Woman on Earth. *Harper's Magazine,* August.
Red Eden. *The Believer,* May.
PAUL ROBERTS
The Last Drops. *Popular Science,* July.
ADAM RUBEN
Experimental Error: Fetus Don't Fail Me Now. sciencecareers.sciencemag.org, May 27.

HAMPTON SIDES
The Man Who Saw Too Much. *Outside,* January.
CHARLES SIEBERT
Joy for Orphan Elephants. *National Geographic,* September.
ERIK SOFGE
The Early Adopter's Guide to Space Travel. *Popular Mechanics,* May.
DON STAP
Grains of Change. *Audubon,* March–April.
PAUL J. STEINHARDT
The Inflation Debate. *Scientific American,* April.
NEAL STEPHENSON
Innovation Starvation. www.wired.com, October 27.
DAWN STOVER
The "Scientization" of Yucca Mountain. www.thebulletin.org, October 12.
MEERA SUBRAMANIAN
India's Vanishing Vultures. *Virginia Quarterly Review,* Spring.
ELIZABETH SVOBODA
Is Fusion Finally for Real? *Popular Mechanics,* June.

STEPHAN TALTY
Shark! *Playboy,* June.
GARY TAUBES
Is Sugar Toxic? *New York Times,* April 13.
DICK TERESI
Lynn Margulis Says She's Not Controversial, She's Right. *Discover,* April.
ABIGAIL TUCKER
Dig, Drink and Be Merry. *Smithsonian,* August.

ERIK VANCE
Dolphin Desperadoes. *Discover,* September.
VLATKA VEDRAL
Living in a Quantum World. *Scientific American,* June.

KATY VINE
The Birdman of Texas. *Texas Monthly,* May.
JAMES VLAHOS
Blue Sky Power. *Popular Mechanics,* March.

BARRY YEOMAN
Lake Erie Deathwatch. www.onearth.org, August 31.

JOCELYN C. ZUCKERMAN
Constant Gardeners. *OnEarth,* Winter.
MARLENE ZUK
Six Legs. *Natural History,* June.